THE QUANTUM AGE

HOW THE PHYSICS OF THE VERY SMALL HAS TRANSFORMED OUR LIVES

BRIAN CLEGG

ICON

This edition published in the UK in 2015 by
Icon Books Ltd, Omnibus Business Centre,
39–41 North Road, London N7 9DP
email: info@iconbooks.com
www.iconbooks.com

First published in the UK in 2014 by Icon Books Ltd

Sold in the UK, Europe and Asia
by Faber & Faber Ltd, Bloomsbury House,
74–77 Great Russell Street,
London WC1B 3DA or their agents

Distributed in the UK, Europe and Asia
by TBS Ltd, TBS Distribution Centre, Colchester Road,
Frating Green, Colchester CO7 7DW

Distributed in South Africa
by Jonathan Ball, Office B4, The District,
41 Sir Lowry Road, Woodstock 7925

Distributed in Australia and New Zealand
by Allen & Unwin Pty Ltd,
PO Box 8500, 83 Alexander Street,
Crows Nest, NSW 2065

Distributed in Canada by
Publishers Group Canada,
76 Stafford Street, Unit 300,
Toronto, Ontario M6J 2S1

Distributed to the trade in the USA
by Consortium Book Sales and Distribution
The Keg House, 34 Thirteenth Avenue NE, Suite 101
Minneapolis, Minnesota 55413-1007

ISBN: 978-184831-846-5

Typeset in Melior by Marie Doherty

Printed and bound in the UK by
Clays Ltd, St Ives plc

Contents

Praise for *The Quantum Age*

'If you are looking for an enjoyable read into all things
quantum physics and how it is applied to everyday life,
look no further.'

BBC Focus

'Brian Clegg does a superb job of explaining complicated
scientific concepts in easily understood language. *The
Quantum Age* is his best book yet, because the concepts
he explains are central to our everyday lives in the
21st century, even though most people think they are
incomprehensible and abstruse. From how the Sun keeps
shining to the quantum computer revolution there is plenty
here to enthral and entertain, as well as to inform.'

John Gribbin, author of *In Search of Schrödinger's Cat*
and *Computing With Quantum Cats*

'What sets this book apart is the way it focuses on the
applications of quantum physics – the things that have
changed our lives and brought us to what Clegg calls the
"quantum age". Truly fascinating.'

Times Higher Education

'I challenge anyone not to find it spellbinding.'

Nick Smith, *E&T Magazine*

For Gillian, Chelsea and Rebecca

Acknowledgements

With thanks as always to my editor, Duncan Heath, for his help and support, and to all those who have provided me with information and assistance – you know who you are.

One person I would like to mention by name is the late Richard Feynman, whose books enthralled me and who turned quantum theory from a confusing mystery to an exciting challenge.

About the author

Science writer **Brian Clegg** studied physics at Cambridge University and specialises in making the strangest aspects of the universe – from infinity to time travel and quantum theory – accessible to the general reader. He is editor of www.popularscience.co.uk and a Fellow of the Royal Society of Arts. His previous books include *Science for Life*, *Inflight Science*, *Build Your Own Time Machine*, *The Universe Inside You*, *Dice World* and *Introducing Infinity: A Graphic Guide*.

www.brianclegg.net

Introduction

The chances are that most of the time you were at school your science teachers lied to you. Much of the science, and specifically the physics, they taught you was rooted in the Victorian age (which is quite probably why so many people find school science dull). Quantum theory, special and general relativity, arguably the most significant fundamentals of physics, were developed in the 20th century and yet these are largely ignored in schools, in part because they are considered too 'difficult' and in part because many of the teachers have little idea about these subjects themselves. And that's a terrible pity, when you consider that in terms of impact on your everyday life, one of these two subjects is quite possibly the most important bit of scientific knowledge there is.

Relativity is fascinating and often truly mind-boggling, but with the exception of gravity, which I admit is rather useful, it has few applications that influence our experience. GPS satellites have to be corrected for both special and general relativity, but that's about it, because the 'classical' physics that predates Einstein's work is a very close approximation to what's observed unless you travel at close to the speed of light, and is good enough to deal with everything from the acceleration of a car to planning a Moon launch. But quantum physics is entirely different. While it too is fascinating and mind-boggling, it also lies behind everything. All the objects we see and touch and use are made up of quantum particles. As is the light we use to see those objects. As are you. As is the Sun and all the other stars. What's more, the process that fuels the Sun, nuclear fusion, depends on quantum physics to work.

That makes the subject interesting in its own right, something you really should have studied at school; but there is far more, because quantum science doesn't just underlie the basic building blocks of physics: it is there in everyday practical applications all around you. It has been estimated that around 35 per cent of GDP in advanced countries comes from technology that makes use of quantum physics in an active fashion, not just in the atoms that make it up. This has not always been the case – we have undergone a revolution that just hasn't been given an appropriate label yet.

This is not the first time that human beings have experienced major changes in the way they live as a result of the development of technology. Historians often highlight this by devising a technological 'age'. So, for instance, we had the stone, bronze and iron ages as these newly workable materials made it possible to produce more versatile and effective tools and products. In the 19th century we entered the steam age, when applied thermodynamics transformed our ability to produce power, moving us from depending on the basic effort of animals and the unpredictable force of wind and water to the controlled might of steam. And though it is yet to be formally recognised as such, we are now in the quantum age.

It isn't entirely clear when this era began. It is possible to argue that the use of current electricity was the first use of true quantum technology, as the flow of electricity through conductors is a quantum process, though of course none of the electrical pioneers were aware that this was the case. If that is a little too concealed a usage to be a revolution, then there can be no doubt that the introduction of electronics, a technology that makes conscious use of quantum effects, meant that we had moved into a new phase of the world. Since then we have piled on all sorts of explicitly quantum devices from the ubiquitous laser to the MRI scanner. Every time we use a mobile phone, watch TV, use a supermarket checkout or take a photograph we are making use of sophisticated quantum effects.

Without quantum physics there would be no matter, no light, no Sun ... and most important, no iPhones.

I've already used the word 'quantum' thirteen times, not counting the title pages and cover. So it makes sense to begin by getting a feel for what this 'quantum' word means and to explore the weird and wonderful science that lies behind it.

CHAPTER 1

Enter the quantum

Until the 20th century it was assumed that matter was much the same on whatever scale you looked at it. When back in Ancient Greek times a group of philosophers imagined what would happen if you cut something up into smaller and smaller pieces until you reached a piece that was uncuttable (*atomos*), they envisaged that atoms would be just smaller versions of what we observe. A cheese atom, for instance, would be no different, except in scale, to a block of cheese. But quantum theory turned our view on its head. As we explore the world of the very small, such as photons of light, electrons and our modern understanding of atoms, they behave like nothing we can directly experience with our senses.

A paradigm shift

Realising the very different reality at the quantum level was what historians of science like to give the pompous term a 'paradigm shift'. Suddenly, the way that scientists looked at the world became different. Before the quantum revolution it was assumed that atoms (if they existed at all – many scientists didn't really believe in them before the 20th century) were just like tiny little balls of the stuff they made up. Quantum physics showed that they behaved so weirdly that an atom of, say, carbon has to be treated as if it is something totally different to a piece of graphite or diamond – and yet all that is inside that lump of graphite or diamond is a collection of these carbon atoms. The behaviour of quantum particles is strange indeed, but that does not mean that it is unapproachable without a doctorate in physics. I quite happily teach the basics of quantum theory to ten-year-olds. Not the maths, but you don't need mathematics to appreciate what's going on. You just need the ability to suspend your disbelief. Because quantum particles refuse to behave the way you'd expect.

As the great 20th-century quantum physicist Richard Feynman (we'll meet him again in detail before long) said in a public lecture: '[Y]ou think I'm going to explain it to you so you can understand it? No, you're not going to be able to understand it. Why, then, am I going to bother you with all this? Why are you going to sit here all this time, when you won't be able to understand what I am going to say? It is my task to persuade you *not* to turn away because you don't understand it. You see, my physics students don't understand it either. This is because *I* don't understand it. Nobody does.'

It might seem that Feynman had found a good way to turn off his audience before he had started by telling them that they wouldn't understand his talk. And surely it's ridiculous for me to suggest I can teach this stuff to ten-year-olds when the great Feynman said he didn't understand it? But he went on to explain what he meant. It's not that his audience wouldn't be able to understand what took place, what quantum physics described. It's just that no one knows *why* it happens the way it does. And because what it does defies common sense, this can cause us problems. In fact quantum theory is arguably easier for ten-year-olds to accept than adults, which is one of the reasons I think that it (and relativity) should be taught in junior school. But that's the subject of a different book.

As Feynman went on to say: 'I'm going to describe to you how Nature is – and if you don't like it, that's going to get in the way of your understanding it … The theory of quantum electrodynamics [the theory governing the interaction of light and matter] describes Nature as absurd from the point of view of common sense. And it agrees fully with experiment. So I hope you can accept Nature as she is – absurd.' We need to accept and embrace the viewpoint of an unlikely enthusiast for the subject, the novelist D.H. Lawrence, who commented that he liked quantum theory *because* he didn't understand it.

The shock of the new

Part of the reason that quantum physics proved such a shocking, seismic shift is that around the start of the 20th century, scientists were, to be honest, rather smug about their level of understanding – an attitude they had probably never had before, and certainly should never have had since (though you can see it creeping in with some modern scientists). The hubris of the scientific establishment is probably best summed up by the words of a leading physicist of the time, William Thomson, Lord Kelvin. In 1900 he commented, no doubt in rounded, self-satisfied tones: 'There is nothing new to be discovered in physics. All that remains is more and more precise measurement.' As a remark that he would come to bitterly regret this is surely up there with the famous clanger of Thomas J. Watson Snr, who as chairman of IBM made the impressively non-prophetic remark in 1943: 'I think there is a world market for maybe five computers.'

Within months of Kelvin's pronouncement, his certainty was being undermined by a German physicist called Max Planck. Planck was trying to iron out a small irritant to Kelvin's supposed 'nothing new' – a technical problem that was given the impressive nickname 'the ultraviolet catastrophe'. We have all seen how things give off light when they are heated up. For instance, take a piece of iron and put it in a furnace and it will first glow red, then yellow, before getting to white heat that will become tinged with blue. The 'catastrophe' that the physics of the day predicted was that the power of the light emitted by a hot body should be proportional to the square of the frequency of that light. This meant that even at room temperature, everything should be glowing blue and blasting out even more ultraviolet light. This was both evidently not happening and impossible.

To fix the problem, Planck cheated. He imagined that light could *not* be given off in whatever-sized amounts you like, as you would expect if it were a wave. Waves could come in any size or wavelength – they were infinitely variable, rather than being broken into discrete components. (And everyone knew that light was a wave, just as you were taught at school in the Victorian science we still impose on our children.)

Instead, Planck thought, what if the light could come out only in fixed-sized chunks? This sorted out the problem. Limit light to chunks and plug it into the maths and you didn't get the runaway effect. Planck was very clear – he didn't think light actually *did* come in chunks (or 'quanta' as he called them, the plural of the Latin *quantum* which roughly means 'how much'), but it was a useful trick to make the maths work. Why this was the case, he had no idea, as he *knew* that light was a wave because there were plenty of experiments to prove it.

Mr Young's experiment

Perhaps the best-known example of these experiments, and one we will come back to a number of times, is Young's slits, the masterpiece of polymath Thomas Young (1773–1829). This well-off medical doctor and amateur scientist was obviously remarkable from an early age. He taught himself to read when he was two, something his parents discovered only when he asked for help with some of the longer words in the Bible. By the time he was thirteen he was a fluent reader in Greek, Latin, Hebrew, Italian and French. This was a natural precursor to one of Young's impressive claims to fame – he made the first partial translation of Egyptian hieroglyphs. But his language abilities don't reflect the breadth of his interests, from discovering the concept of elasticity in engineering to producing mortality tables to help insurance companies set their premiums.

His big breakthrough in understanding light came while studying the effect of temperature on the formation of dewdrops – there really was nothing in nature that didn't interest this man. While watching the effect of candlelight on a fine mist of water droplets he discovered that they produced a series of coloured rings when the light then fell on a white screen. Young suspected that this effect was caused by interactions between waves of light, proving the wave nature that Christiaan Huygens had proclaimed back in Newton's time. By 1801, Young was ready to prove this with an experiment that has been the definitive demonstration that light is a wave ever since.

Young produced a sharp beam of light using a slit in a piece of card and shone this light onto two parallel slits, close together in another piece of card, finally letting the result fall on a screen behind. You might expect that each slit would project a bright line on the screen, but what Young observed was a series of alternating dark and light bands. To Young this was clear evidence that light was a wave. The waves from the two slits were interfering with each other. When the side-to-side ripples in both waves were heading in the same direction – say both up – at the point they met the screen, the result was a bright band. If the wave ripples were heading in opposite directions, one up and one down, they would cancel each other out and produce a dark band. A similar effect can be spotted if you drop two stones into still water near to each other and watch how the ripples interact – some waves reinforce, some cancel out. It is natural wave behaviour.

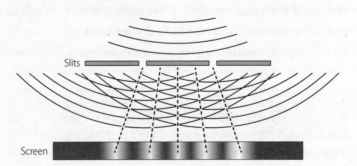

Dotted lines show where waves reinforce,
producing bright patches on screen

Fig. 1. Young's slits.

It was this kind of demonstration that persuaded Planck that his quanta were nothing more than a work-around to make the calculations match what was observed, because light simply *had* to be a wave – but he was to be proved wrong by a man who was less worried about convention than the older Planck, Albert Einstein. Einstein was to show that Planck's idea was far closer to reality than Planck would ever accept. This discrepancy in viewpoint was glaringly obvious when Planck recommended Einstein for the Prussian Academy of Sciences in 1913. Planck requested the academy to overlook the fact that Einstein sometimes 'missed the target in his speculations, as for example, in his theory of light quanta …'.

The Einstein touch

That 'speculation' was made by Einstein in 1905 when he was a young man of 26 (forget the white-haired icon we all know: this was a dapper young man-about-town). For Einstein, 1905 was a remarkable year in which the budding scientist, who was yet to achieve a doctorate and was technically an amateur, came up with the concept of special relativity,* showed how Brownian motion† could be explained, making it clear that atoms really did exist,

---

* Einstein's expansion of Galileo's theory of relativity. Galileo had observed that all movement has to be measured relative to something, but Einstein added that light always travels at the same speed. This special relativity shows that time and space are linked and dependent on the observer's motion.

† The observation by the Scottish botanist Robert Brown (1773–1858) that pollen grains suspended in water danced around. Einstein showed how this could be caused by fast-moving water molecules colliding with the grains.

and devised an explanation for the photoelectric effect (see page 13) that turned Planck's useful calculating method into a model of reality.

Einstein was never one to worry too much about fitting expectations. As a boy he struggled with the rigid nature of German schooling, getting himself a reputation for being lazy and uncooperative. By the time he was sixteen, when most students had little more on their mind than getting through their exams and getting on with the opposite sex, he decided that he could no longer tolerate being a German citizen. (Not that young Albert was the classic geek in finding it difficult to get on with the girls – quite the reverse.) Hoping to become a Swiss citizen, Einstein applied to the exclusive Federal Institute of Technology, the Eidgenössische Technische Hochschule or ETH, in Zürich. Certain of his own abilities in the sciences, Einstein took the entrance exam – and failed.

His problem was a combination of youth and very tightly focused interests. Einstein had not seen the point of spending much time on subjects outside the sciences, but the ETH examination was designed to pick out all-rounders. However, the principal of the school was impressed by young Albert and recommended he spent a year in a Swiss secondary school to gain a more appropriate education. Next year, Einstein applied again and got through. The ETH certainly allowed Einstein more flexibility to follow his dreams than the rigid German schools, though his headstrong approach made the head of the physics department, Heinrich Weber, comment to his student: 'You're a very clever boy, but you have one big fault. You will never allow yourself to be told anything.'

After graduating, Einstein tried to get a post by writing to famous scientists, asking them to take him on as an assistant. When this unlikely strategy failed, he took a position as a teacher, primarily to be able to gain Swiss citizenship, as he had already renounced his German nationality, so was technically stateless. Soon, though, he would get another job, one that would give him plenty of time to think. Einstein successfully applied for the post of Patent Officer (third class) in the Swiss Patent Office in Bern.

Electricity from light

It was while working there in 1905 that Einstein turned Planck's useful trick into the real foundation of quantum theory, writing the paper that would win him the Nobel Prize. The subject was the photoelectric effect, the science behind the solar cells we see all over the place these days producing electricity from sunlight. By the early 1900s, scientists and engineers were well aware of this effect, although at the time it was studied only in metals, rather than the semiconductors that have made modern photoelectric cells viable. That the photoelectric effect occurred was no big surprise. It was known that light had an electrical component, so it seemed reasonable that it might be able to give a push to electrons* in a piece of metal and produce a small current. But there was something odd about the way this happened.

---

* The electron is the negatively charged fundamental particle that occupies the outer reaches of atoms and carries electrical current.

A couple of years earlier, the Hungarian Philipp Lenard had experimented widely with the effect and found that it didn't matter how bright the light was that was shone on the metal – the electrons freed from the metal by light of a particular colour always had the same energy. If you moved down the spectrum of light, you would eventually reach a colour where no electrons flowed at all, however bright the light was. But this didn't make any sense if light was a wave. It was as if the sea could only wash something away if the waves came very frequently, while vast, towering waves with a low frequency could not move a single grain of sand.

Einstein realised that Planck's quanta, his imaginary packets of light, would provide an explanation. If light were made up of a series of particles, rather than a wave, it would produce the effects that were seen. An individual particle of light* could knock out an electron only if it had enough energy to do so, and as far as light was concerned, higher energy corresponded to being further up the spectrum. But the outcome had no connection with the number of photons present – the brightness of the light – as the effect was produced by an interaction between a single photon and an electron.

Einstein had not only turned Planck's useful mathematical cheat into a description of reality and explained the photoelectric effect, he had set the foundation for the whole of quantum physics, a theory that, ironically, he would spend much of his working life challenging. In less than a decade, Einstein's concept of the 'real' quantum

---

* They wouldn't be known as photons until the 1920s when they were given the name by the American chemist Gilbert Lewis.

would be picked up by the young Danish physicist Niels Bohr to explain a serious problem with the atom. Because atoms really shouldn't be stable.

Uncuttable matter

As we have seen, the idea of atoms goes all the way back to the Ancient Greeks. It was picked up by British chemist John Dalton (1766–1844) as an explanation for the nature of elements, but it was only in the early 20th century (encouraged by another of Einstein's 1905 papers, the one on Brownian motion) that the concept of the atom was taken seriously as a real thing, rather than a metaphorical concept. The original idea of an atom was that it was the ultimate division of matter – that Greek word for uncuttable, *atomos* – but the British physicist Joseph John Thomson (usually known as J.J.) had discovered in 1897 that atoms could give off a smaller particle he called an electron, which seemed to be the same whatever kind of atom produced it. He deduced that the electron was a component of atoms – that atoms were cuttable after all.

The electron is negatively charged, while atoms have no electrical charge, so there had to be something else in there, something positive to balance it out. Thomson dreamed up what would become known as the 'plum pudding model' of the atom. In this, a collection of electrons (the plums in the pudding) are suspended in a sort of jelly of positive charge. Originally Thomson thought that all the mass of the atom came from the electrons – which meant that even the lightest atom, hydrogen, should contain well over a thousand electrons – but later work suggested that there was mass in the positive part of the atom too, and hydrogen, for example, had only the single electron we recognise today.

Bohr's voyage of discovery

When 25-year-old physicist Niels Bohr won a scholarship to spend a year studying atoms away from his native Denmark he had no doubt where he wanted to go – to work on atoms with the great Thomson. And so in 1911 he came to Cambridge, armed with a copy of Dickens' *The Pickwick Papers* and a dictionary in an attempt to improve his limited English. Unfortunately he got off to a bad start by telling Thomson at their first meeting that a calculation in one of the great man's books was wrong. Rather than collaborating with Thomson as he had imagined, Bohr hardly saw the then star of Cambridge physics, spending most of his time allocated to his least favourite activity, undertaking experiments.

Towards the end of 1911, though, two chance meetings changed Bohr's future and paved the way for the development of quantum theory. First, on a visit to a family friend in Manchester, and again at a ten-course dinner in Cambridge, Bohr met the imposing New Zealand physicist Ernest Rutherford, then working at Manchester University. Rutherford had recently overthrown the plum pudding model by showing that most of the atom's mass was concentrated in a positive-charged lump occupying a tiny nucleus at the heart of the atom. Rutherford seemed a much more attractive person to work for than Thomson, and Bohr was soon heading for Manchester.

There Bohr put together his first ideas that would form the basis of the quantum atom. It might seem natural to assume that an atom with a (relatively) massive nucleus and a collection of smaller electrons on the outside was similar in form to a solar system, with the gravitational force that keeps the planets in place replaced by the electromagnetic attraction between the positively charged nucleus and the negatively charged electrons. But despite the fact that this picture is still often employed to illustrate the atom (it's almost impossible to restrain illustrators from using it), it incorporates a fundamental problem. If an electron were to orbit around the nucleus it would spurt out energy and collapse into the centre, because an accelerating electrical charge gives off energy – and to keep in orbit, an electron would have to accelerate. Yet it was no better imagining that the electrons were fixed in position. There was no stable configuration where the electrons didn't move. This presented Bohr with a huge challenge.

Inspired by discovering reports of experiments showing that when heated, atoms gave off light photons of distinct energies, Bohr suggested something radical. Yes, he decided, the electrons could be in orbits – but only if those orbits were fixed, more like railway tracks than the freely variable orbit of a satellite. And to move between two tracks required a fixed amount of energy, corresponding to absorbing or giving off a photon. Not only was light 'quantised', so was the structure of the atom. An electron could not drift from level to level, it could only jump from one distinct orbit to another.

Inside the atom

An atom is an amazing thing, so it is worth spending a moment thinking about what it appears to be like. That traditional picture of a solar system is still a useful starting point, despite the fatal flaw. To begin with, just like a solar system, the atom has a massive bit at the centre and much less massive bits on the outside. If we look at the simplest atom, hydrogen, it has a single positively charged particle – a proton – as a nucleus and a single negatively charged electron outside of it. The proton, the nucleus, is nearly 2,000 times more massive than the electron, just as the Sun is much more massive than the Earth. And like a solar system, an atom is mostly made up of empty space.

One of the earliest and still most effective illustrations of the amount of emptiness in an atom is that if you imagine the nucleus of an atom to be the size of a fly, the whole atom will be about the size of a cathedral – and apart from the vague presence of the electron(s) on the outside, all the rest is empty space. Now we need to move away from the solar system model, though. I've already mentioned that a true solar-system-style atom would collapse. Another difference is that, unlike the solar system, the electrons and the nucleus are attracted by electromagnetism rather than gravity. And here we come across a real oddity, with a Nobel Prize waiting for anyone who can explain it. The electron has exactly the same magnitude of charge (if opposite in value) to the positive charge on a proton in the nucleus. No one has a clue why, but it's rather handy in making atoms work the way they do. The solar system has no equivalent to this. Gravity comes in only one flavour.

The final reason we have to throw away the solar system model is that electrons simply don't travel around nuclei in nice, well-defined orbits, the same way that planets travel around the Sun. They don't even move around on the sort of rail tracks that Bohr first envisaged. As we will discover, quantum particles are never so considerate and predictable as to do something like this. A better picture of an electron is a sort of fuzzy cloud of probability spread around the outside of the atom, rather than those sweeping orbit lines so favoured by graphic designers – though that is much harder to draw. More on that in a moment.

Building on Bohr

It would be an exaggeration to say that Bohr's idea for the structure of atoms transformed our view of physics on its own – apart from anything, his original model worked only for the simplest atom, hydrogen. But before long a group of young physicists – with de Broglie, Heisenberg, Schrödinger and Dirac to the fore – had picked up the baton and were pushing forward to build quantum theory into an effective description of the way that atoms and other quantum particles like photons behave. And their message was that they behave very badly indeed – at least if we expect them to carry on the way we expect ordinary everyday objects to act.

Louis de Broglie showed that Einstein's transformation of the wavy nature of light into particles was a two-way street – because quantum objects we usually thought of as particles, like atoms and electrons, could just as happily behave as if they were waves. It was even possible to do a variant of the two-slit experiment with particles, producing interference patterns. Werner Heisenberg, meanwhile, was uncomfortable with Bohr's orbits modelled on the 'real' observed world and totally abandoned the idea of trying to provide an explanation of quantum particles that could be envisaged. He developed a purely mathematical method of predicting the behaviour of quantum particles called matrix mechanics. The matrices (two-dimensional arrays of numbers) did not represent anything directly observable – they were simply values that, when manipulated the right way, produced the same results as were seen in nature.

Erwin Schrödinger, always more comfortable than Heisenberg with something that could be visualised, came up with an alternative formulation known as wave mechanics that it was initially hoped described the behaviour of de Broglie's waves. Paul Dirac would eventually show that Schrödinger's and Heisenberg's approaches were entirely equivalent. But Schrödinger was mistaken if he believed he had tamed the quantum wildness. If his wave equation had truly described the behaviour of particles it would show that quantum particles gradually spread out over time, becoming immense. This was absurd. To make matters worse, the solutions of his wave equations contained imaginary numbers, which generally indicated there was something wrong with the maths.

Numbers that can't be real

Imaginary numbers had been around as a concept since the 16th century. They were based on the idea of square roots. As you probably remember from school, the square root of a number is the value which, multiplied by itself, produces that original number. So, for instance, the square root of 4 is 2. Or, rather, 2 is *one* of 4's square roots. Because it is also true that −2 multiplied by itself makes 4. The number 4 has two square roots, 2 and −2. But this leaves a bit of a gap in the square root landscape. What, for example, is the square root of −4? It can't be 2, nor can it be −2, as both of those produce 4 when multiplied by themselves. So what can the square root of a negative number be? To deal with this, mathematicians invented an arbitrary value for the square root of −1, referred to as 'i'. Once i exists, we can say the square roots of −4 are 2i and −2i. These numbers based on i are imaginary numbers.

This would seem to be the kind of thing mathematicians do in their spare time to amuse themselves − quite entertaining, but of no interest in the real world. But in fact complex numbers, which have both a real and an imaginary component, such as 3+2i, proved to be very useful in physics and engineering. This is because by representing a complex number as a point plotted on a graph, where the real numbers are on the x axis and the imaginary numbers on the y axis, a complex number provides a single value that represents a point in two dimensions. As long as the imaginary parts cancel out before coming up with a real world prediction, complex numbers proved a great tool. But in Schrödinger's wave equation, the imaginary numbers did not politely go away, staying around to the embarrassment of all concerned.

Probability on the square

This mess was sorted out by Einstein's good friend, Max Born. Born worked out that Schrödinger's equation did not actually say how a particle like an electron or a photon behaved. Instead of showing the location of a particle, it showed the *probability* of a particle being in a particular location. To be more precise, the square of the equation showed the probability, handily disposing of those inconvenient imaginary numbers. Where it was inconceivable that the particle itself would spread out over time, it was perfectly reasonable that the *probability* of finding it in any location would spread out this way. But the price that was paid for Born's fix was that probability became a central part of our description of reality. Born's explanation of the equation worked wonderfully, though it had to be taken on trust – no one could say why, for instance, it was necessary to square the outcome.

There is nothing new in using probability to describe a level of uncertainty. I can demonstrate this if I put a dog in the middle of a park and close my eyes for ten seconds. I don't know exactly where that dog will be when I open my eyes. I can say, though, that it will probably be within about 20 metres of where I left it, and the probability is higher that it will be near the lamppost than that it will be halfway up a beech tree or taking a ride on the roundabout. However, this use of probability in the ordinary world does not reflect reality, but rather the uncertainty in my knowledge. The dog will actually be in a particular location at all times with 100 per cent certainty – I just don't know what that location is until I open my eyes.

If instead of a dog I was observing a quantum particle, Schrödinger's equation, newly explained by Born, also gives me the probability of finding the particle in the different possible locations available to it. But the difference here is that there is no underlying reality of which I am unaware before I look. Until I make the measurement and produce a location for the particle, the probability is all that existed. The particle wasn't 'really' in the place I eventually found it up until the point the measurement was made.

Taking this viewpoint requires a huge stretch of the imagination (which is probably why ten-year-olds cope with quantum theory better than grown-ups), but if you can overcome common sense's attempt to put you straight, it throws away the problems we face when thinking, for instance, of how the Young's slits experiment could possibly work with photons of light. If you remember, the traditional wave picture had waves passing through both slits and interfering with each other to create the pattern of fringes on the screen. But how could this work with photons (or electrons)? This difficulty is made particularly poignant if you consider that we can now fine-tune the production of these particles to the extent that they can be sent towards the slits one at a time – and yet still, over time, the interference pattern, caused by the interaction of waves of probability, builds up on the screen.

Where is that particle?

There is a very dangerous temptation that almost all science communicators fall into at this point. I have to admit I have done it frequently in the past. And I have heard TV scientist Brian Cox do it too, commenting on his radio show *The Infinite Monkey Cage* that the photon is in two places at once. In fact Cox's book, *The Quantum Universe* (co-authored with Jeff Forshaw), even has a chapter entitled 'Being in two places at once'. The tempting but faulty description is that quantum theory says that a photon can be in two places at once, so it manages to go through both slits and interferes with itself. However, this gives a misleading picture of what is really happening in the probabilistic world of the quantum.

What would be much more accurate would be to say that a photon in the Young's slits experiment isn't *anywhere* until it hits the screen and is registered. Up to that point all that exists is a series of probabilities for its location, described by the (square of the) wave equation. As these waves of probability encompass both slits, then the final result at the screen is that those probability waves interfere – but the waves are not the photon itself. If the experimenter puts a detector in one of the slits that lets a photon through but detects its passing, the interference pattern disappears. We have forced the photon to have a location and there is no opportunity for the probability waves to interfere.

It was this fundamental role for probability that so irritated Einstein, making him write several times to Max Born that this idea simply couldn't be right, as God did not play dice. As Einstein put it, when describing one of the quantum effects that are controlled by probability: 'In that case, I would rather be a cobbler, or even an employee in a gaming house, than a physicist.'

It was from the central role of probability that Heisenberg would deduce the famous Uncertainty Principle. He showed that quantum particles have pairs of properties – location and momentum, for instance, or energy and time – that are intimately related by probability. The more accurately you discover one of these pairs of values, the less accurately it is possible to know the other. If, for instance, you knew the exact momentum (mass times velocity) of a particle, then it could be located anywhere in the universe.

The infernal cat

It is probably necessary also at this point to mention Schrödinger's cat, not because it gives us any great insights into quantum theory, but rather because it is so often mentioned when quantum physics comes up that it needs putting into context. This thought experiment was dreamed up by Schrödinger to demonstrate how absurd he felt the probabilistic nature of quantum theory became when it was linked to the 'macro' world that we observe every day.

In the Young's slits experiment, even single photons produce an interference pattern as described above – but if you check which slit a photon goes through, the probabilities collapse into an actual value and the pattern disappears. Quantum particles typically get into 'superposed' states until they are observed. (Superposition just says that a particle has simultaneous probabilities of being in a range of states, rather than having an actual unique state.) In the cat experiment, a quantum particle of a radioactive material is used to trigger the release of a deadly gas when the particle decays. The gas then kills a cat that is in a box. Because the radioactive particle is a quantum particle, until observed it is in a superposed state, merely a combination of the probabilities of it being decayed or not decayed. Which presumably leaves the cat in a superposed state of alive and dead. Which is more than a little weird.

In reality, the moggy doesn't seem to have much to worry about, at least as far as being superposed goes – it can, of course, still die. As the experiment is described, it is assumed that the particle, and hence the cat, is in a superposed state until the box is opened. Yet in the Young's slit experiment the mere presence of a detector is enough to collapse the states and produce an actual value for which slit the particle travelled through. So there is no reason to assume that the detector in the cat experiment that triggers the gas would not also collapse the states. But Schrödinger's cat is such a favourite with science writers – if only because it gives illustrators something interesting to draw – that it really needs highlighting.

Because it is so famous, the cat has a tendency to turn up in other quantum thought experiments. The original Schrödinger's cat experiment is all about the fuzzy borderline between the quantum world of the very small and the classical world we observe around us. Experimenters are always trying to stretch that boundary, achieving superposition and other quantum effects for larger and larger objects. Until recently there was no good measure of just what 'bigger' meant in this context – how to measure how macroscopic or microscopic (and liable to quantum effects) an object was. However, in 2013 Stefan Nimmrichter and Klaus Hornberger of the University of Duisburg-Essen devised a mathematical measure that describes the minimum modification required in the appropriate Schrödinger's equation to destroy a quantum state, giving a numerical measure of just how realistic a superposition would be.

This measure produces a value that compares any

26

given superposition with a single electron's ability to stay in a superposed state. For example, the biggest molecule that has been superposed to date has 356 atoms. The theorists calculated that this would have a 'macroscopicity' factor of 12, which means it being superposed for a second is on a par with an electron staying superposed for 10^{12} seconds. There is reasonable expectation that items with a factor of up to around 23 could be put into a superposed state. To put this into context, and in honour of Schrödinger, the theorists also calculated the macroscopicity of a cat.

They started with a classic physicist's simplification by assuming that the cat was a 4-kilogram sphere of water, and that it managed to get into a superposition of being in two places 10 centimetres apart for one second. The result of the calculation was a factor of around 57 – it was the equivalent of putting an electron into a superposed state for 10^{57} seconds, around 10^{39} times the age of the universe, stressing just how unlikely this is – though it is worth noting that even the 10^{23} expectation is longer than the lifetime of the universe. Unlikely things do happen (if rather infrequently), and quantum researchers are always careful never to say 'never'.

It is these weird aspects of quantum theory that make the field so counterintuitive … and so fascinating. And nowhere more so than when quantum effects crop up in the natural world. Quantum theory is not just something that is relevant to the lab, or even to high-tech engineering. It has a direct impact on the world around us, from the operation of the Sun that is so central to life on Earth, to some of the more subtle aspects of biology.

CHAPTER 2

Quantum nature

Because of the way we are taught science, it is tempting to divide the subject up into tight compartments. Physics, for instance, is about how stuff behaves, while biology explains the living side of nature. (As someone with a physics background, I might cruelly say that chemistry is the clean-up operation for the bits in between that neither of the other subjects wants.) But these labels and divisions are arbitrary and human-imposed. Quantum theory has no intention of staying confined in the box labelled physics. Nature makes use of quantum processes.

It's quantum all the way down

At a fundamental level, this is a truism about nature. Given that atoms and light are governed by quantum theory, and pretty well everything in nature is either atoms or light,* it is inevitable that quantum processes rule. Quantum physics describes why atoms exist and why they don't collapse. So you could say that when you watch a rabbit run across a field or examine the beautiful structure of an orchid you are seeing a product of quantum

---

* Purists will point out that it is not true that pretty well every-thing is atoms and light. After all, around 68 per cent of the universe is dark energy, and 27 per cent dark matter. But the remaining 5 per cent, the bit we can actually experience – because the rest is, by definition, unobserved – is predominantly atoms and light.

theory. But that's just the foundation level, explaining the component parts of nature. Quantum theory also applies at a far higher level than the basics of how atoms work.

Perhaps the most dramatic example of this is the Sun. Because it is so far away and seems little more than a bright light in the sky, we tend to underestimate the significance of the Sun to life on Earth. This hasn't always been the case. Earlier civilisations worshipped the Sun as a god for a good reason. Being closer to the land, they were aware of the Sun's significance in helping their crops grow. And without artificial lights, they had a lot to thank the Sun for in enabling them to see. In a modern world, where we are rarely far from a light at night, whether it's in our home, a street light or the glow from our phones, it is hard to appreciate just how dark and scary the natural world at night can be. Sit for a while in a pitch-dark cave, ideally with the howling of wolves thrown in for full impact, and you can see why the Sun's contribution during the day was so appreciated.

Even our ancestors, though, underestimated the importance of the Sun. Just imagine there was no Sun, that the Earth was a lone planet, wandering through space. What would we miss out on? There would be no weather – weather is powered by the Sun, producing temperature differences to create wind and evaporating water to generate clouds and rain. Temperatures on the Earth would drop to below −250°C. There never would have been an oxygen atmosphere, as there would be no photosynthesis. But all this is irrelevant in a sense, because there would be no Earth. Without the Sun's gravitational pull, the material that came together to make up the Earth would have remained scattered in space. We owe our existence to the Sun.

How old is the Sun?

When those observing the Sun got past simply regarding it as a light in the sky, they typically considered it to be a fire. After all, what else glows like that? But the idea of a heavenly bonfire was itself a problem, because we all know that fires don't burn for ever. This was a real problem when it became obvious in Victorian times that the Earth had been around far longer than suggested by the traditional creation date, worked out from adding up Bible 'begats' back to 4004 BC. Two factors were responsible for this. One was geology. By observing the way erosion acts at the present, geologists were able to estimate that the natural formations we see must have faced erosion for hundreds of millions of years. The other Victorian bugbear for ageing the Sun was evolution. Darwin made it clear that the kind of processes he had in mind for evolution by natural selection would also require hundreds of millions of years for species to evolve.

Set against these long timescales were the physicists, trying to come up with an explanation for how the Sun worked, notably William Thomson, later Lord Kelvin. Kelvin had first considered the possibility that the Sun was simply burning, but if it were coal – which sounds silly now, but was seriously considered then – it would last only a few thousand years, and even with the best energy/weight reaction available, that of hydrogen and oxygen, it could at best have a lifetime of 20,000 years. That was far too short for any sensible model of what was observed on the Earth. And it was ridiculous that the Earth should be far older than the Sun. Kelvin also considered whether the Sun could be externally heated by the impact of meteors in collision with it. But he reckoned that to achieve the output it did, it would require about two Earths'-worth of matter to hit it every century. This steady increase in mass should have produced significant modifications to the orbits of the planets, which had not been observed. This left only one possibility that Kelvin could think of.

He suggested that the Sun had formed by a cloud of gas coming together through the pull of gravity. As the atoms were compressed closer together, this would raise the temperature. Think of what happens when you repeatedly push the handle of a bicycle pump – it warms up. He suggested this compression heating happened to such a degree that the Sun became intensely hot as it formed, and it had then spent its life radiating away that heat, like a piece of iron heated in a forge, which continues to glow and give off heat long after it is taken out of the flames. Kelvin calculated that with the immense mass of the Sun it could continue to radiate (though it would be gradually getting dimmer) for around 30 million years.

In a way, Kelvin's idea was very clever. We do still think that this contraction under the pull of gravity, creating heat and pressure, is how stars like the Sun form – but it is not how they keep burning. However, that 30 million-year timescale put Kelvin in direct contradiction with the geologists and Darwin. (The mild-mannered Darwin, rather than challenge Kelvin, was horrified by the potential conflict, and removed references to the duration of evolution in editions of *Origin of Species* from this point on. Privately he referred to Kelvin as an 'odious spectre'.) The result was something of an impasse. The only theories for the formation of the Earth required the Sun to be there first. The only way the Sun could work put its age at less than 30 million years, yet more and more evidence on the Earth suggested that it had been in existence for many hundreds of millions of years. In fact we now know it's around 4.5 billion years old.

The power of fusion

The solution to this conundrum was the discovery of a new way that the Sun could be powered. Under immense temperature and pressure, hydrogen ions could fuse to produce the next heaviest element, helium. In this process, energy is given off. Scale this up to the size of the Sun and you have a body that is capable of producing energy for billions of years – the Sun seems to be around halfway through a 10 billion-year lifespan. The very process of nuclear fusion is a quantum process, providing a central role for quantum theory in the existence of Earth, humans and all of nature, but there is one more quantum card to be played in the explanation of the Sun's workings. Because, while fusion power is certainly a likely process to produce the amount of energy required, it turned out that even the heat and pressure in the heart of the Sun is not enough to make fusion happen.

To produce helium requires four protons – each a hydrogen ion, the nucleus of a hydrogen atom with its electron stripped away – to come together in close proximity. As it happens, the actual process is a little more complex, with an isotope of helium, helium–3, forming, then pairs of the helium coming together to produce the stable helium–4 and release a pair of protons. But the essential outcome is that four hydrogen ions have become a helium ion and produced energy along the way. Those hydrogen ions, protons, are positively charged, so they repel each other. The closer they get, the stronger that repulsion becomes. They can fuse only when the strong nuclear force takes over. But this has no effect outside very short distances, so the protons have to get ridiculously close to each other.

Here is where the weirdness of quantum physics comes in. Bear in mind that Schrödinger's equation tells us that, over time, the possible locations of a particle spread out. So though two protons are most likely to be kept too far away from each other to fuse, there is a small probability that they *are* close enough already – and in those small number of cases, fusion takes place. Another way to look at this, which gives the process its name, is to think of the electromagnetic repulsion as a barrier, keeping the protons apart. A few protons will undergo a process called quantum mechanical tunnelling, which means they appear on the other side of the barrier without passing through the space in between. They jump to the other side and fuse.

Although this tunnelling has a low probability, there are so many protons in the Sun that several millions of tons of them fuse every second. And all because of this strange quantum effect. Without tunnelling, the Sun's fusion reaction would not work. The Sun would not be giving off the energy that it does. Meaning no weather, no oxygen and impossibly low temperatures on the Earth.

There would be no life without this uniquely quantum process.

Enzyme enablers

More down to Earth, we are discovering an increasing range of quantum processes cropping up in nature, where they might not have been expected before. A well-established example that was uncovered in the 1970s is the work of enzymes as catalysts. Enzymes are large organic molecules, usually proteins, that take part in chemical reactions within living things, including the human body. For example, enzymes help with the digestion of our food, acting as catalysts to hugely speed up chemical reactions that otherwise would be too slow to support life.

Catalysts work by making chemical reactions work more easily, reducing the amount of energy needed for the reaction, but the catalyst is not part of the final output of the reaction, so it is freed up to be used again. A catalyst might, for instance, change the nature of a chemical bond or combine with one compound to produce an intermediate substance that is much more reactive. In the action of some enzymes, either a proton or an electron undergoes tunnelling, just like the protons in the Sun. Without the tunnelling, only those protons or electrons with enough energy to get over the barrier preventing the reaction taking place would succeed. It's not that a new reaction takes place because of the quantum effect, but rather the enzyme produces a much faster reaction than expected, typically thousands of times faster. Without this quantum boost many biological organisms – including humans – would be unable to function.

It's all in the DNA

Another place where quantum tunnelling seems to take place is in a hugely important bit of the biochemistry of every living thing: DNA mutation. As it's hard to avoid knowing, DNA (short for deoxyribonucleic acid) is the family of molecules that carries our genetic information. It is the instruction set by which we are constructed, and passes on our genes to our offspring. When a cell splits into two so an organism can grow, the DNA itself splits down the middle, unzipping the spiral staircase of its structure to produce two halves. These aren't identical, but each complements the other.

Each 'tread' of the DNA spiral staircase consists of two organic compounds selected from a set of four, known as bases: cytosine, guanine, adenine and thymine. These always pair the same way: cytosine with guanine and adenine with thymine. (If you think of them as the capital letters usually used to represent them, C, G, A and T, the letters made of curves pair, and those made of straight lines pair.) So it is easy to reproduce the missing half of the DNA from the bases in the available one.

The bond that links together the base pairs until the DNA is 'unzipped' is called a hydrogen bond. It's the same kind of bond that links water molecules together, giving water an unexpectedly high boiling point. Hydrogen bonding is an electrical effect, where the relatively positive part of one molecule is attracted to the relatively negative part of another molecule. In the case of water, the positive part is hydrogen, which has only a single negative electron that is tied up in the bond to the rest of the water molecule, and the negative part is oxygen.

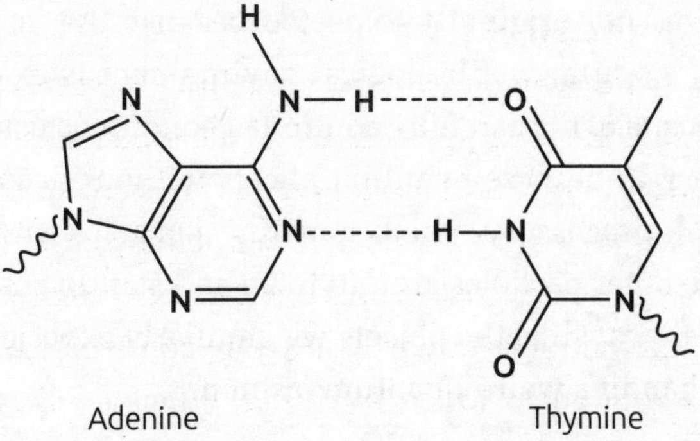

Fig. 2. A-T base pair showing hydrogen bonds (dotted).

In DNA there are hydrogen bonds linking each pair of bases. Each pair has a hydrogen nucleus – a single proton – at one side of the bond. This proton is a quantum particle and that means it can tunnel, in this case across to the other side of the bond so it becomes a part of the other compound. So, for instance, in a pair where A is linked to T, a hydrogen nucleus from the A side could tunnel through to the T while a hydrogen from the T tunnels across the second bond to the A. The formulae for the two bases remain the same, but the structure is now different. And that means that when the DNA unzips, the variant of A is changed enough in shape to bond with C instead of T. The new copy of DNA that results will be a mutation that could result in a change in the organism for which it controls development.

This process has not been experimentally confirmed yet, though it is believed to be a strong possibility for the mechanism behind this kind of mutation. And if it is, it means that a specifically quantum process directly causes changes in living cells. This sort of high-level quantum effect was not originally expected because the 'messi-ness' of a warm, wet biological environment is exactly the opposite of the carefully controlled conditions usually necessary to observe quantum effects without decoher-ence, the process by which quantum particles interact with the other particles around them and start to act in a 'classical' way, like the objects we familiarly experience, rather than in a weird quantum fashion.

Plant light

One of the most dramatic and important biological pro-cesses that is likely to involve high-level quantum effects is photosynthesis, the process by which plants convert light into energy. As we will see later on, any interaction between light and matter is quantum mechanical, just as anything involving an atom or electron is, but recent stud-ies of photosynthesis have shown that quantum physics probably has a more functional role.

Even without any quantum mechanical weirdness, photo-synthesis is a marvel of natural technology. The first hint of something remarkable going on with plants when exposed to light was an accidental discovery by Joseph Priestley. In the mid-1770s, this trouble-prone Non-conformist minister had settled in the unlikely job of librarian to the Earl of Shelburne in his stately home, Bowood House. In return for Priestley's company and conversation, Shelburne was prepared to support Priestley's curiosity on the nature of air and its components. Arguably as a form of entertainment to show off to visiting guests, Shelburne gave over a small room next to his library to Priestley, where the scientist could undertake his experiments.

Priestley is usually credited as the discoverer of oxygen, though he himself would not have recognised its existence, as he was a supporter of the phlogiston theory. This proposed that there was a part of matter called phlogiston that was given off as the matter burned. Air could hold only so much phlogiston, so if, for instance, you put a candle in a bell jar it would go out when the air became fully phlogisticated. In reality this was because the oxygen in the air was being consumed – phlogiston was a sort of anti-oxygen. Priestley discovered that a mouse inside the bell jar would also make the air phlogisticated, causing the mouse to keel over, but if he put a green plant in with the mouse, it would seem to restore this 'injured air' and keep the animal alive.

This awareness that plants somehow repaired something that was restricted or limited when something was burned or an animal breathed was as far as Priestley got, but towards the end of the 18th century, French pastor Jean Senebier and the Swiss scientist Theodore de Saussure showed that the 'injured air' was carbon dioxide, produced by burning or respiration, and that plants could convert this gas to oxygen and carbon-based molecules under the influence of light. We now know that photosynthesis from the Sun effectively feeds the Earth, acting directly on green plants and particularly algae, which use up more than half of the solar energy going to photosynthesis, and indirectly providing that energy needed by the animals that eat the plants (or eat the animals that eat the plants) in our complex food chain.

The physics and chemistry involved in photosynthesis is convoluted, with a whole chain of reactions taking place. First the light pushes up the energy levels of electrons in special coloured molecules like the green chlorophyll in a plant. This energy is converted to chemical form by the photosynthetic reaction centre, which produces oxygen and incorporates carbon into the plant. One of the steps of this intricate process is the fastest known chemical reaction in existence, taking place in a trillionth of a second.

The oddities of the quantum world come into play in the energy's journey from that first excitation of an electron in chlorophyll to its arrival in the reaction centre, where it gets to work converting carbon dioxide to sugars (and releasing some oxygen in the bargain). The way the energy passes from molecule to molecule on the way in is a result of quantum particles behaving like waves. The energised wave of the first excited electron extends into the next molecule, passing on the excitation, and so on. What's more, these waves don't seem to take a random drunkard's walk, but rather they overlap, coming into coherence – the state where the waves all ripple together that makes a laser work (see page 129).

This coherent behaviour had been postulated for a while, and there was some weak evidence of it existing from large plant samples, but in 2013 researchers in Spain and Glasgow discovered it at the molecular level, training lasers on single light-processing molecules to observe the detailed workings of the reaction centres that convert photons to chemical energy. Experiments also on light-harvesting purple bacteria showed that the ability of the quantum particle to probabilistically explore all routes and find the best path meant that the connections change as parts of the organism move, constantly tuning the process, meaning that the conversion can reach levels of around 90 per cent efficiency, far higher than a solar cell (and possibly with implications for photovoltaic cell development in the future).

The pigeon's compass

A rather less certain but fascinating possibility is that a quantum effect is behind one of the marvels of nature – the way that birds like homing pigeons can navigate, apparently by picking up the Earth's magnetic field, using a built-in compass. This mysterious ability has been linked to magnetic particles in their beaks, but there is also evidence that may be stronger that the process is triggered by light hitting the retina of the bird's eye. (In fact three mechanisms have been proposed, and it is entirely possible pigeons use some combination of them.)

When light hits the receptor in the bird's eye, it is used to split a molecule to form two free radicals. These are very reactive molecules that have an unpaired electron (it's free radicals that are restrained by antioxidants from causing cell damage). These electrons can act as tiny magnetic compasses, with their quantum property called spin influenced by a magnetic field. Typically one radical will be closer to the nearest atomic nucleus, and hence feels the magnetic field less than the other. This difference between the two gives the chemicals a different level of reactivity, making it possible for the bird to get some feedback from the interaction, perhaps by the synthesis of a chemical in the retina. The two unpaired electrons are created entangled, linked to each other in a quantum fashion, and this could help amplify the effect.

I think, therefore I'm quantum

The most extreme – and most contentious – overlap between quantum theory and biology is the idea that consciousness itself is a quantum phenomenon. Although there is no direct evidence to base this theory on, some have suggested that it is not possible to explain the phenomenon of the conscious mind using conventional classical physics, and that it needs quantum effects like entanglement to make it possible. One suggestion, with the clumsy name of 'orchestrated objective reduction', comes from physicist Roger Penrose and medical doctor Stuart Hameroff.

Penrose proposed that the brain is capable of computation that would be impossible using conventional mechanisms, with the probabilistic nature at the heart of quantum theory explaining this extra capability. Hameroff, an anaesthetist, suggested that the cytoskeleton, the structure that supports the neurons in the brain, and in particular microtubules, thin polymers that form part of the cytoskeleton, could act as quantum systems, where electrons tunnel between the microtubules.

The idea that consciousness involves quantum effects does not seem to stretch the bounds of probability to too great an extent, though as yet the jury is out. We just don't understand what consciousness is, or the mechanism behind it, well enough to explore how much it could depend on quantum effects. However, what certainly can be done is to make use of the mathematics behind quantum theory to get a better understanding of some aspects of human behaviour.

Quantum voting

Andrei Khrennikov of Linnaeus University, Sweden and Emmanuel Haven at the University of Leicester took the maths used to describe quantum decoherence and applied it to the American political system. Specifically they looked at the voters' choice between Republican and Democrat in the combination of presidential and congressional elections. Their idea was that the mental state of a voter could be considered a superposition of the states 'Democrat' and 'Republican' with a certain probability for each, and they treated the two elections as if they were entangled qubits (the processing units of quantum computers – see page 231). This gave them a vehicle for exploring the dynamics of the voters' mental states when exposed to media information.

It's early days but there is some evidence that this kind of tool could be valuable in gaining insights into the way we think and make decisions. Even if this approach proves effective, it isn't evidence that the decision-making itself is dependent on quantum physics. Most likely it is just that the maths happens to work well as a model because both the political situation and the state of quantum particles involve probabilistic measurements of properties that can exist only in a small number of states, and that have to 'collapse' into a single fixed value, in the case of politics when a vote is cast. But it is another way that the power of quantum physics can give us insights into biological processes.

Biology is a good example of a field that grew from simple observation to a true science, able to explain what was happening in nature. This involved building a growing knowledge of the detail of what happens within a biological structure, detail that would eventually be discovered to include quantum effects. Another field that is purely quantum but began with simple observations is the realm of electricity. It's time to sing the body electric.

CHAPTER 3

The electron's realm

In a lecture hall in the Royal Society in London, lit by the flickering glow of oil lamps, a strange performance is under way. It looks like a kind of ritual undertaken by a secret society in a display of bizarre debauchery. A boy is suspended from the ceiling by silk ropes. He reaches out to touch a girl who stands just in reach on a tar-topped barrel. She, in turn, holds out her hand and a stream of feathers float upwards from a table towards her fingers, as if magically attracted to her.

The philosopher's amber

What the assembled members were witnessing was a popular scientific demonstration of the 18th century known as 'the electrical boy'. The boy's feet were picking up a charge from a hand-cranked device that generated static electricity. Exactly what electricity was, no one was sure. But clearly it was *something* that could pass straight through the boy on its way to give the girl the ability to levitate the pile of feathers. It was typically produced by rubbing a rotating disc or sphere, often made of glass, with a suitable cloth like wool, though in previous times amber had been employed to generate the effect, which gives us the word 'electricity' from the Greek for amber, *elektron*. Without any clear explanation of why this was happening, it was thought that this action produced some sort of invisible fluid that could pass through the body.

It might seem silly to consider electricity as something that behaves like water – after all, it doesn't exactly run out of the socket if we leave the toaster unplugged. Yet we do still happily use terms (including plug) that are more appropriate to a fluid – we speak of an electrical current, for instance, just like the current of a river.

As with any major development in scientific understanding there were lots of people along the way who contributed to our picture of electricity and its cousin, magnetism. We could dally with the likes of Ampère and Oersted but to pick out the biggest milestones along the way, it really will take only three men to lead us from the baffled amazement produced by the quaint demonstrations of the 18th century up to the 20th-century realisation that electricity was a quantum phenomenon. The first of these was Michael Faraday.

The wizard of Albemarle Street

Faraday stands out in many ways. These days we would expect a physicist to have a university education and an excellent grasp of mathematics. Faraday had neither. Back then, to be a scientist usually meant being a rich dilettante. Faraday started practically penniless. As the son of a blacksmith who had taken the major step of moving from Westmorland to London in search of work, Michael Faraday probably thought himself lucky at the age of fourteen in 1805 to be apprenticed to a bookbinder. He was learning a trade that should eventually make him a decent income, and he had the opportunity to attend a self-improvement group called the City Philosophical Society, which would be the key to transforming his life.

Faraday took careful notes of the lectures he attended at the Society, and was allowed by his employer to bind them, producing a leather-backed volume that so impressed the bookbinder that he showed the notes to a rich client who was one of the many who enjoyed visits to the Royal Institution. This was a then brand-new establishment in Albemarle Street, just off fashionable Piccadilly, that both promoted research and put on lectures to encourage the public understanding of science. The client, a Mr Dance, gave Faraday tickets to attend the RI's top-of-the-bill performer – Humphry Davy, the ultimate Victorian scientific celebrity. This seems to have inspired Faraday so much that with Mr Dance's help he got the opportunity to stand in as Davy's secretary when the great man was temporarily blinded in an accident. Then, after a return to the bookbinder's, Faraday became Davy's lab assistant and general factotum after the incumbent was sacked for drunkenness.

Wollaston's folly

Faraday quickly settled into the Royal Institution and made remarkable progress, given his lack of education. By 1821 he had received a promotion, got married and moved into the rooms at the Institution previously occupied by Davy. Everything seemed rosy – yet his first big discovery in the field of electricity and magnetism threatened to ruin his fairy-tale success. Davy had asked Faraday to write up the current state of knowledge on electricity and magnetism, but Faraday was always a hands-on person, and rather than simply describe the experimental results he had seen reported, he reproduced the experiments himself. At one point he had set up an electrical current running through a wire next to a magnet. The wire began to move around the magnet in circles. This was not in the literature he was summarising.

Understandably excited, Faraday rushed to publish his results – only to be accused of stealing the discovery from an elder statesman of the Royal Institution, William Wollaston. Wollaston had developed a theory that described electricity moving along wires in a corkscrew spiral motion. He had asked Davy for help to find evidence of this, but nothing ensued. Now here was Faraday, claiming to have discovered a circling motion associated with electricity. Clearly Wollaston thought – though with no basis in fact, as there was no link to his incorrect theory – that Faraday had stolen his idea.

The implication of cheating horrified Faraday, who had deeply held religious beliefs. He turned to his old mentor Humphry Davy for the expected support, knowing that Wollaston's theory bore no resemblance to his experimental results. But Davy sided with his friend and social equal Wollaston, rather than supporting the working-class Faraday. The social divide took priority over scientific fact. This was the end of any friendship between the two. When, for instance, Faraday was elected to the Royal Society just one person opposed him – Davy. Yet the scientific community were very clear that the discovery was Faraday's. Not only had he done original work unconnected to Wollaston's theory, he had provided the basis for the electric motor.

Despite this support in the wider world, Faraday largely abandoned electricity and magnetism for ten years as a result of the unpleasantness, concentrating on chemistry and taking on an administrative role that enabled him to start the regular 9.00pm Friday Discourses (these days something of a pantomime, as the audience is still expected to turn up in black tie) and the Christmas Lectures for Children, which in later televised form would be the inspiration for many a young 20th-century British scientist. But the appeal of electricity and magnetism did not go away. By 1831 Faraday was tempted back to the field by hearing of experiments where an electrical current in one wire seemed magically to produce a second current in another wire, despite being at a distance.

The new generation

Once he set up a pair of coils of wire and connected up the first of them to produce a current, Faraday expected to see the other start to produce a continuous flow of electricity, but instead there was a brief surge in the second coil when he switched the first on or off. He knew that magnetism could act at a distance, and that an electrical coil could act as a magnet. From this he made the leap to deducing that it was a changing level of magnetism from the first coil that produced electricity in the second. He was soon able to reproduce this effect by moving an ordinary permanent magnet through a coil, inventing the generator.

It was about this time that the often mentioned apocryphal conversation with Prime Minister Robert Peel is said to have taken place, when the politician asked Faraday what use his invention was, to get the reply: 'I know not, but I wager one day your government will tax it.'

Faraday had one more essential contribution to make in this story, though we will meet him again in a later chapter. A more traditional physicist – and certainly any modern physicist – would have attempted to explain what was happening through the medium of mathematics. But Faraday was no mathematician, arguably the last physicist who could make major discoveries without it. He had seen how magnets pull iron filings sprinkled on a sheet of paper above them into lines linking the magnet's two poles. In the dim, gaslit laboratory, Faraday imagined these lines glowing around a magnet.

When he moved a wire near a magnet or pushed a magnet through a coil, it was as if the wire was cutting through the lines of force, as he termed them. The closer the lines were together, the more the wire cut them, the more current was generated. This model worked well with the business of switching on or off an electromagnet. When it was switched on, the lines of force expanded out from the magnet, cutting themselves on the wire. When switched off, the lines collapsed and the reverse happened.

This way of imagining electromagnetic interactions as a field of force would eventually become hugely important in physics, not just in understanding electromagnetism but at the heart of quantum theory itself. But as yet, there was still more to discover about the nature of electricity and magnetism. The person who would take Faraday's elegant but innumerate concepts and turn them into the first modern scientific view was James Clerk Maxwell.

A Scottish savant

Maxwell was born in Edinburgh in 1831, a member of a later generation than Faraday and with a very different background from the older man. Some have argued (rather weakly) that Maxwell could be called the first scientist, in part because the word 'scientist' was coined only in 1834, holding the same relationship to science as an artist did to art. Until then, clumsy terms like 'natural philosopher' or 'savant' were likely to be used. This may be tenuous, but what certainly is true is that Maxwell could be called the first *modern* scientist, as he pioneered scientific theories that were driven by mathematics, an approach that would have been totally alien to Faraday.

Where Faraday was put to work as soon as he was able, Maxwell had a free and enviable childhood – at least in his early years. He was allowed to explore and experiment in his family's country house, Glenlair, on the estate at Middlebie in Galloway, dabbling with everything from crystals to hot air balloons. This idyll was shattered when his mother died. Though a private tutor was first tried for the eight-year-old Maxwell, he was soon sent to Edinburgh Academy, which must have seemed like a descent into hell after the freedom he had experienced.

Maxwell was smaller than average, had a stutter and a country accent, and was more interested in his books and experiments than sports. Some children thrive at boarding schools, but Maxwell, known to his classmates as Dafty, was the classic target for school bullies. He had to endure this until the age of sixteen when he had the blessed release of moving to Edinburgh University, and three years later to Cambridge. The recommendation to the Master of Trinity College, from Professor James Forbes of Edinburgh, described him in mixed fashion: 'He is not a little uncouth in his manners, but withal one of the most original young men I have ever met with.'

After his graduation in 1854, Maxwell wanted to follow in the footsteps of his personal hero, Michael Faraday. He would work on a wide range of subjects, producing, among other things, the first colour photograph; but the aspect of his work that he will always be remembered for was his mathematical summary of the relationship between electricity and magnetism discovered by Faraday. Talk to a physicist about elegant equations that simply capture a description of the world and many will point to Maxwell's equations – originally eight in total, but simplified by Oliver Heavyside and Heinrich Hertz to produce four neat, short equations that remain at the heart of our understanding of the universe.

Thomson's tiny bodies

The final part of our triumvirate is the Mancunian scientist J.J. Thomson, who would later fail to get on with Niels Bohr. Thomson joined Owens College (later Manchester University) at the age of fourteen, and six years later moved on to Cambridge, where he remained for the rest of his career. Thomson's driving interest was the structure of the atom (which was why it was so sad that he didn't get on with Bohr – see page 16), but he worked widely in electricity and magnetism, and in 1897 his studies of cathode rays would give him the opportunity to reach for fame (and a Nobel Prize).

Cathode rays were first observed by Michael Faraday (him again) in the 1830s when he passed a current through a glass tube with reduced air pressure inside and saw a bright glow in the tube, but they were properly studied only once it was possible to remove the majority of the air from a tube, notably by British scientist William Crookes, which resulted in these tubes being known as Crookes tubes. Something seemed to travel down the tube between the two electrodes from the negative cathode to the positive anode, which was often shaped like a Maltese cross. Whatever was travelling seemed to be moving fast enough to sometimes overshoot the anode and crash into the glass at the end of the tube, causing it to give off a distinctive green glow.

Just what was travelling down the tubes was the subject of considerable debate – they were called 'cathode rays' because they were emitted from the negatively charged cathode. Crookes himself developed the theory that they were charged atoms from the residual air in the tube, while others, including Heinrich Hertz, thought that they were a new kind of electromagnetic wave – a variant of light. Thomson, however, proved them both wrong when he succeeded in measuring the mass of the carrier particles in these invisible rays and found that this was non-zero – and so the cathode rays were not light – but discovered that they had only a tiny fraction of the mass of atoms. What's more, they were identical in charge and mass to other similar electrically charged products like that from the photoelectric effect.

Thomson concluded that 'the carriers of negative electricity are bodies, which I have called corpuscles, having a mass much smaller than that of the atom of any known element, and are of the same character from whatever source the negative electricity may be derived'. Thomson's corpuscles were what would become known as electrons, the name given to them by George Stoney a few years later. It was soon realised that what Thomson had found was not just the content of cathode rays but the constituents of the current in all conventional electrical circuits. Electrons flowed through metal wires, just as they passed through the evacuated tubes.

There was a slight embarrassment with this discovery, as it meant that the traditional way of representing electrical current in circuit diagrams, flowing from positive to negative, was back to front, but there wasn't a lot that could be done about this. Given our practical experience of electricity, you might think that electrons gush through wires at extremely high speeds – a good fraction of the speed of light. After all, when we flick a switch, we don't have to wait for the electricity to get to the other end as we would with water travelling through a pipe. But electrical conduction is rather more complex than a wire providing a conduit for a gushing current of electrons.

A conductor's life

The most familiar electrical conduction is through a metal. The structure of the metal is an array of atoms like a latticework with the outer electrons of the metal relatively free, able to detach themselves from their parent atoms and float about in the lattice. Usually their movement is random, floating about until they collide with something as a result of thermal energy, but if an electric field is applied to the wire by making one end negative and the other positive, the electrons drift towards the positive pole. This movement is surprisingly leisurely – they often cover only around a metre per second. In effect the electrons travel at around walking pace.

It might seem then that when we switch on an electrical current it should take hours to pass down a long enough wire, which would be true if the wire started out empty and had to gradually fill up with electrons. But in reality, the wire already contains electrons along its length. When the circuit is made, the electrical field is carried along it by an electromagnetic wave, travelling at the speed of light in the material. This means that electrons start to move throughout the wire at almost the same time, so there is no need to wait for them to get from one end to the other.

Electrons are, of course, quantum particles exhibiting all the weirdness of quantum behaviour, so anything involving electricity is inherently a quantum process. If we think of conduction in a metal, we also see the quantum structure of the atom coming into play in what is known as band structure. In a single copper atom, say, the extension of Bohr's work on the hydrogen atom tells us that there will be a series of 'orbitals', fixed levels of energy that an electron can occupy, while it is not allowed to be in the levels in between. As more and more atoms are brought together to form the complex structure of the solid metal, something interesting happens.

While the inner electrons remain associated with their own atoms, in orbitals belonging to a single atom, the outer electrons can take on shared orbitals that run across atoms in the metal. As more atoms are added in, there are more and more orbitals available, squeezed closer and closer together until the gaps between them are negligible. They form a band, a continuous range that enables the electrons to move freely within the metal. These free electrons carry heat – which is why metals are good conductors of heat – and electricity.

From electricity to electronics

The first electronic devices made use of the basic behaviour of electrons. Resistors, for instance, which reduce the flow of electrons through a circuit, are usually made of a mix of conducting and insulating materials, reducing the ease with which electrons can flow. Another definitive aspect of electronics – the ability to control and switch the flow of current that enables us to produce the logic gates necessary for computers – is based on the ability to control the direction of flow of electrons and to switch one current using another. In the early days of electronics, these tasks were handled by thermionic valves (known as vacuum tubes in the US), which were like Crookes tubes but more functional (and significantly smaller).

The best example of the ability of one current to control another, the essential switching role at the heart of computing, would be in the triode valve. Here we have what amounts to a traditional Crookes tube – a glass tube with most of the air removed and with a cathode and an anode, which are two pieces of conductor inserted through the wall of the tube, so that electrons can flow from the cathode to the anode. The cathode would normally be heated in a valve, hence the device's typical glow (and the heat it gave off), to give the electrons extra energy, making it easier for them to flow freely. The switching ability came from another electrode in the form of a grid that sat in the path of the flow of electrons. If this grid was given a negative charge, it repelled the electrons and stopped them passing through, switching off the flow through the tube.

As well as acting as a simple on/off switch, the triode could also function as an amplifier. A small amount of current applied to the grid could control a much bigger current flowing through the valve. So the small variations in the grid current were amplified into much bigger variations in the main current. If, for instance, the grid had an alternating current with a complex waveform applied to it, the main current would replicate that waveform but with a bigger amplitude, enabling radios and music players to boost the relatively weak signal picked up from radio waves or from a recording, where it would typically be produced by a needle in a gramophone record pushing on a crystal that produced electricity when twisted, a process known as piezoelectricity.

Valves worked – and are still occasionally used, as some people believe they produce a particularly warm and attractive sound in the reproduction of audio (though blind testing suggests a lot of this is an audio placebo effect, where enthusiasts hear what they want to hear). But devices based on valves weren't without their problems. For a start, the glass tubes were fragile – easily damaged and anything but safely portable. They were also relatively large, the smallest being around the size of a thumb and some, where a major current had to be handled, bigger than a whole hand. What's more, the need for a heater meant that, like a light bulb, they would eat up energy and would eventually burn out and need replacing.

Computing with electrons

Building something as complex as a computer using valves was a job on an industrial scale. The first programmable electronic computer, Colossus, used at Bletchley Park during the Second World War to help decode German messages, had 1,500 valves in the original version, and 2,400 in the Mark 2. By the time the US built ENIAC, a step up on Colossus in being truly general purpose, the valve count had risen to over 17,000. These massive machines bore no resemblance to computing as we now know it, which was why it wasn't entirely stupid when Thomas Watson, the head of IBM, remarked that the world probably had a demand for about five computers. There was never going to be a market for many ENIACs.

ENIAC weighed around 27 tonnes, was 30 metres (100 feet) long and consumed 150 kilowatts of electricity. The vast majority of this electrical power went to heat, so this monstrous device was pumping out enough warmth to require the building to be constantly cooled (leading to a whole generation of computer rooms with special air conditioning). Because of the inevitable and regular failure of the valves, the longest ENIAC ever ran without breaking down was just under five days, and a more typical time between failure was two days. We might moan about misbehaving modern computers, but they are astonishingly reliable by comparison.

Although a fair number of valve-based computers were built, originally for military and university use and later for businesses, they remained a tiny corner of the growing applications of electronics in everyday life. Before long practically every house had a 'wireless' – a radio that used valves, giving it the characteristic 'warm up' time after the equipment was switched on. But less than ten years from ENIAC's first use, as early as 1954, the transistor started to take over from the valve, both in domestic electronics and computing.

We will come back to how it works, but a transistor managed to do the same job as the triode valve, controlling one electrical current with another, but using a solid chunk of material. There was no heater (hence no need to warm up), no need for a delicate glass enclosure, no vacuum to maintain. And transistors could be made much smaller than the equivalent valve – many were smaller than a fingernail. The first transistor was invented by John Bardeen, William Shockley and Walter Brattain, making the trio rare physics Nobel Prize winners for the invention of a piece of technology, rather than a scientific discovery.

The first computer to use transistors was built at the University of Manchester in 1953 – its 92 transistors a vanishingly small count when compared with the 10^{19} transistors (that's 1 with 19 zeroes after it) turned out each year now, but it was the start of a transformation of the electronics industry, moving production from expensive hand-crafted manufacturing to cheap industrial-scale mass production.

Integration rules

The last step in the development of modern electronics was the move to integrated circuits. The early radios, computers and other electronic gear based on solid-state electronics consisted of individual components like transistors, resistors and capacitors soldered onto printed circuit boards. These boards were simply a sheet of plastic where the wiring between the components was replaced by lines in a metal film on the surface of the plastic. The shape of the circuit was etched onto the metal by starting with a board coated with a complete metal film and painting the parts to be kept with a resisting chemical, then dipping the board into acid that ate away the unprotected metal.

Even in the 1960s, consumer electronics and computers (then still very much an industrial tool) were dependent on these circuit boards with their masses of individual components. It meant that simple devices like a radio could be made much smaller than they had been before, able to be carried around in the hand or fitted in the dashboard of a car. But computers, requiring many thousands of transistors, would still need hundreds of circuit boards, so they had to be accommodated in room-sized cabinets and needed serious cooling systems. The mainframe computers of the 1960s may have been smaller than ENIAC – and packed in significantly more power – but they were not exactly miniature or suitable for the home.

To produce the kinds of electronics we are familiar with today, where computers became a desktop box, or a tablet, or even a pocket-sized computer in the form of a smartphone (such phones are in reality a powerful computer with added features like the radio transmitters required for calls and Bluetooth), integrated circuits had to be employed. Devised in the late 1950s and becoming practically available from the mid-1960s, these put all the elements of an electronic circuit – the transistors, resistors and so on – onto the surface of a single chip of silicon.

A hopeless conductor

Both solid-state transistors and the later integrated circuits were made possible as a result of the discovery of the flexibility of what otherwise might have seemed a useless type of substance. Many materials are either conductors like metals, which electricity flows through easily, or insulators like ceramics, which prevent electricity from flowing at all, both valuable in electrical circuits. But there is a third class of material, semiconductors, which allow limited conduction, often under the influence of a secondary input, that would make all these solid-state electronics possible.

The operation of semiconductors is a purely quantum effect that, unlike ordinary electricity, could not even be understood without a grasp of quantum theory. Valves were quantum devices, but they could be built and operated without realising what was going on, using a simple model like the control of a flow of water. With the introduction of transistors, we saw the first technology emerging that required an understanding of quantum physics to design it. The electronic device, originally a hybrid that used quantum processes but could be (mis)understood in a classical fashion, was dropping its classical heritage and going entirely weird.

If we go back to the idea of energy bands in conductors, an insulator has a big gap (known by the inspired name 'band gap') between the 'valence bands' where the bound electrons stay with the atoms and the conduction bands where electrons can move freely. This means it is difficult for an electron to get free. In a semiconductor that gap is narrower, but the substance still acts primarily as an insulator without some assistance. For some kinds of semiconductor, that kick to get it conducting is provided by light energy. Selenium, for instance, conducts better when light is falling on it. But for the kinds of semiconductors used in transistors and integrated circuits, the boost usually comes from doping, the addition of impurities to the semiconductor.

When current flows in the higher conduction band in a semiconductor, a few of the electrons from the valence band are also kicked up to the conduction band. Due to a complex peculiarity of the way electrons behave, those at the top of the valence band behave oddly and move 'against the current', so they will be flowing in the opposite direction to the electrons in the conduction band, carrying any gaps in the electrons with them. These gaps are known as 'holes' and are treated as if they were particles in their own right. The net result is that electrons flow in one direction in the conduction band and holes move in the opposite direction at the top of the valence band. This process is strongly aided by the doping agents, because they provide an extra band level that has a much smaller gap than usual.

Doping dramatically increases the number of free electrons available to play with. There are two types of doping agent, n for negative and p for positive. An atom of an n-type doping agent adds a spare electron compared to what is available in the original semiconductor, while an atom of a p-type agent has one fewer electron than usual. This may seem a disadvantage, making the semiconductor more like an insulator, but the missing electron provides a hole, effectively operating as a positively charged particle that can move around (in reality electrons are still moving, but as we have seen, the hole moves with them, and this is sometimes easier to deal with mathematically, as a single moving hole is the equivalent of many moving electrons). So, for instance, silicon, the prime semiconductor of modern electronics, might typically be doped with phosphorus to create an n-type semiconductor, or boron for a p-type.

From semiconductor to circuit

A straightforward traditional transistor was often constructed of a sandwich of three sections of semiconductor, corresponding to the three electrodes in a triode valve – these materials would usually be either n, p and n doped materials or p, n, p. By applying a voltage between one side and the central segment of the sandwich, the arrangement of extra band levels means that the small voltage applied acts like a valve, controlling the flow of electrons through from one side of the sandwich to the other.

In an integrated circuit, a more complex arrangement known as a MOSFET (metal oxide semiconductor field effect transistor) is the usual equivalent of that basic transistor, produced by growing a layer of silicon dioxide* on top of the silicon wafer and spraying on a fine layer of metal or a substance known as polycrystalline silicon, producing a more complex layered effect that still produces the essential action of the transistor, but in a much more compact arrangement.

Even working alone, transistors are valuable because the ability of a small varying voltage applied to the central

---

* Also known as silica, the main component of sand or quartz.

segment of the sandwich to control a much higher voltage across from side to side produces amplification. But for a computer, transistors are linked together to form units called logic gates. To see why these are required we need to take a brief step away from electronics and into the Victorian mathematics of Boolean algebra.

Symbols of logic

The man behind this mathematical oddity was George Boole, born in a cobbler's shop in Lincoln in 1815. Although his father was a shoemaker, he had an interest in maths and engineering, teaching young George the basics of mathematics himself. George was educated only to the age of sixteen, never attending university – he went straight to be an assistant schoolmaster in Doncaster, but continued his education by reading, building a considerable expertise in algebraic methods. After running his own school for a while, Boole was appointed to the chair of mathematics at Queen's College, Cork, where he taught for the rest of his life, which is why he is sometimes described as an Irish mathematician.

Five years after taking up the post, Boole published a book on the mathematical theories of logic, which turned logic into a kind of algebra that could manipulate concepts with symbols. As we will see in a moment, his approach is at the heart of the way computers work, but we also use it at a more visible level when specifying a request to a search engine like Google. Imagine that I put in the following request:

(Cars AND trucks) (red OR blue) (NOT Ford)

The words I have highlighted in capitals (which is traditional for Boolean algebra) are effectively the key instructions on how to process the request – they control the search engine. So the 'AND' tells the computer that each result should refer to both cars and trucks. It isn't enough to have just cars or just trucks, both must be present. The middle section uses 'OR', which tells the search engine that as long as the result features either one of red or blue it is fine – it doesn't have to include both, although it can. And the final section tells the search engine I want to exclude Fords from my result. The brackets are just there for clarity of what goes with what. Interestingly, though search engines were originally strictly Boolean, Google appears not to use Boolean controls any more: when I tried this search in Google Images, around half the results were Fords. Google now seems to have a very loose interpretation of 'NOT', possibly overridden by advertising budgets.

Mr Boole's gates

Combinations of simple logical instructions are used to set up all the operations required for the internal workings of a computer that we never see. There, the controls are referred to as 'gates'. So, for instance, an AND gate is one that takes two inputs and returns 1 (represented by an electrical current) if both inputs are 1, while it returns 0 (no current) in any other circumstances. Using Boolean logic, the AND gate returns a result only if both inputs have a value. By contrast, an OR gate returns 1 if either of its inputs has a value. And the NOT gate simply reverses a single input, turning 0 into 1 and 1 into 0. There are also compound gates like a NAND gate that produces the reversed output of an AND gate.

If you had a means to produce these gates electronically, you could assemble them to provide all the main functions of a computer – and transistors (or for that matter valves) do just this. If you put two transistors in series, for instance, the result is an AND gate, because current will flow through only if both transistors are 'open' to the flow of current. If either one is closed off, representing a 0, the current does not flow through the whole structure – effectively putting a 0 out. Only by having both transistors set to 1, making 1 AND 1, do you get an output of 1. Similarly, an OR gate can be produced using two transistors in parallel, so if either is open to allowing a current to flow (i.e. set to 1) then current will flow through the gate, producing an output of 1.

The basic building blocks of electronics can be put together with as much flexibility as Lego bricks, combining to produce everything from an audio amplifier to a computer. When I was young there were popular toy kits that enabled the young scientist to do this, plugging actual components like transistors and resistors into a pegboard circuit. But there is more building on this quantum framework needed to fill out the workings of every device we have produced since. Three pieces of technology in particular demand our attention: memory, screens and digital cameras.

A fading memory

From the earliest days computers have required two kinds of storage – a working memory in which to store bits while they are manipulated by those logic gates, and longer-term storage. Conventional electronic devices were fine for the short-term usage of working memory, but for long-term storage there was the problem that as soon as the valves, or later transistors, lost their power, the memory disappeared. Early computers usually relied on information that was punched as a series of holes in paper tapes or cards, but for much of the lifetime of the digital computer – and still to a considerable extent – the most frequently used forms of long-term storage have been magnetic. Here, information is stored on a metallic surface as a series of magnetic domains – the orientations of small pieces of magnetic materials – originally on drums or on tape, but universally now on the platters of fast-rotating discs.

The 1990s saw a new mode of storage become commonplace – optical storage of data on CD-like technology. But this has proved a surprisingly short-lived phase in the development of computing. Now the long-term storage of choice is often flash memory. This has the advantage of the speed of reading and writing of computer memory – all forms of disc are inevitably slower – without the fragility of a high-speed rotating disc. Anyone like me who has dropped a computer with a hard disc knows that this is not a good move. Because it is solid-state, flash memory can also be made much smaller than a mechanical device, with tens of gigabytes squeezable into a chip the size of a fingernail.

As is usual in electronics, all the means of storage since punched paper and cardboard were phased out have been technically quantum in nature, but just as that quantum mechanism became more explicit in going from valves to solid-state, so the move from magnetised surfaces to flash memory brings quantum phenomena to centre stage in its operation. Flash memory was invented at the Japanese firm Toshiba around the beginning of the 1980s and was originally primarily hidden away and only indirectly accessible, used to hold information that was rarely changed, like the BIOS instructions that a computer uses when it first starts up. This was because the early flash memory was expensive and slow to read and write, so not ideal for everyday rapid memory use.

The first of the new generation of flash memory chips came into use during the 1990s, used in removable memory cards, and these days it provides the storage in phones and compact computers, where we take it for granted that many gigabytes of information can be safely stored on a small, portable device. This type of flash memory is much quicker to access than the earlier version. It does have a limitation – it can't access a single location at a time, pulling in or writing hundreds or thousands of bits simultaneously – but this is easily worked around, is sometimes useful and is always far outweighed by the speed of access.

Flash memory, like the conventional type of storage, makes use of transistors, but these are special variants of MOSFET (see page 63) devices called floating-gate transistors. The gate is the transistor's equivalent of the grid in a triode, the electrode that controls the flow through the valve. In a floating-gate transistor, there are two gates: a traditional 'control' gate, and beneath it, a floating gate which is electrically isolated so it can hold a charge indefinitely, providing the data storage. When the floating gate is charged up, it screens the control gates away from influencing the flow through the transistor, giving it a permanent switching role.

The floating gate is totally isolated by insulators, acting as a screen by induction – which leaves the problem of how to change the value of this inaccessible memory by charging up or erasing the charge on the gate. And it is here that a purely quantum effect takes over. Charge is added to the gate, or wiped from it, by quantum tunnelling (see page 34), the process by which a particle – in this case, an electron – can pass from one side of a barrier to the other without passing through the space in between. Without explicitly exploiting this weird quantum effect, this type of floating-gate transistor could not function. Within any device using flash memory for storage there is a whole bundle of tunnelling going on.

Seeing the data

For a long time the information from computers was presented to the world just as it was stored on paper, first as punched tapes and cards and later as the output of automated typewriters known as teletypes, but there was already a technology for outputting visual information that dated back to Victorian times and that would become the standard for computers and televisions – the cathode ray tube. As we have already seen, cathode rays were really streams of electrons, first discovered when they caused the end of an evacuated glass tube to glow – but it wasn't long before the original experiment was improved on in two specific ways.

The first step forward from the original 'Crookes tube' was to coat the glass at the end of the tube with a phosphor. Glass itself is mildly phosphorescent when hit by a stream of electrons, hence the original ghostly green glow that Crookes and the other experimenters saw, but a phosphor gives off a much brighter light. In a phosphor, the incoming electrons smash into the atoms in the lattice of the material. Some of their kinetic energy is absorbed by the electrons orbiting the atoms of the phosphor, boosting them from the fixed valence band to the higher-level conductance band, where they can drift through the lattice until they reach specially introduced impurities called activators. Captured by the activator, the electron drops down in level and gives off energy in the form of a tiny flash of light – a scintillation.

The early screens worked in black and white, using a mix of zinc cadmium sulphide and zinc sulphide silver, while colour screens have collections of three dots of phosphors, peaking in the blue, green and red regions to produce the primary colours of light. (If you thought the primary colours were red, blue and yellow, as they still teach in primary schools, you were misled. These are simplifications of the secondary colours magenta, cyan and yellow. These 'opposites' of the true primaries are the key colours when using pigments, and as you were probably taught them with paint rather than light itself, you were told a fib to keep things simple.)

With the right phosphors, the glow from a cathode ray screen can be bright and well controlled in colour, but a traditional Crookes tube mechanism simply lights up the whole of the end of the tube (except for the shadow that is cast by the anode). To turn a cathode ray tube into a TV or computer screen it is also necessary to control a tight beam of electrons, moving it across the surface of the screen to paint an image on the phosphor surface.

The image, whether it is writing on a computer screen or a picture on a TV, is built up by rapidly sweeping the beam across the surface of the screen and relying on the phosphor continuing to glow just long enough to still be active when the beam gets back to give it another kick. The direction of the beam is controlled by pairs of electrically charged plates, which steer the stream of electrons as required. Of course, electrons are fired only at the bits of the screen where a glow is required. Gaps are left to leave the black in between. (More precisely it's usually the grey in between. A TV can't show anything darker than the colour of the screen when switched off, but our brain, well experienced at fooling us, turns it into jet black when required for, say, a space scene.)

The really rather crude Victorian technology of the cathode ray tube dominated the way we looked at images and text electronically from the earliest days of TV right through to the 1990s. It was only then that the other types of display began to take over. The trouble with cathode ray tubes is that they were big, bulky and heavy (not to mention needing dangerously high voltages to operate). Because the 'gun' that produced the electrons had to be far enough back from the front of the device to be able to sweep its electron beam across the whole screen, the tubes had to be at least half as deep as they were wide. With the earliest screens, which were just a few inches across, this wasn't much of a problem, but as the image got bigger and bigger, the depth of the tube became a significant embarrassment.

Elegant displays

Enter the flat screen. Flat screens rely on three technologies. The earliest, and still probably the most popular, was LCD – the liquid crystal display. Not only did this do away with the fat backside of a cathode ray tube, it used significantly less energy to produce an image – and was capable of being made much larger. Where the old TVs and monitors made up a picture by controlling the way light is generated from the phosphors, an LCD has a uniform illumination all the time across the screen and instead controls how much light gets through to the viewer using a kind of filter. The secret to the way this works is the liquid crystal itself.

Liquid crystals were first discovered way back in 1888, a strange type of substance that could flow like a liquid but had some of the characteristics of a solid crystal. The particular liquid crystals used in screens have a clever trick up their sleeve. In their natural state they twist light that passes through them. Light has a property called polarisation, which can be thought of as the direction in which the wave of light moves from side to side compared to its direction of travel. (If you prefer to think of light as photons – see the next chapter – then this is just a property of a photon that has a particular direction at right angles to the direction of travel.) When light passes through the liquid crystal, the direction of polarisation is twisted.

In an LCD screen, a large sheet of the crystal is placed between two polarising filters, which are like sieves that only allow through light polarised in a particular direction. The filters are at right angles to each other.

So, for instance, if the back filter works horizontally, only horizontally polarised light gets through. This then hits the front filter, which only lets vertically polarised light through. So it blocks the horizontally polarised light from the back filter. Result – nothing gets through. A dark screen ensues. But put the liquid crystal in between and it rotates the polarisation, turning horizontally polarised light into vertical. This means the light from the illuminated panel behind the horizontal filter shines out of the front of the screen.

So far, so good. But here's the clever twist (literally). When an electrical current is put across the liquid crystal, its molecules, which had been twisted in spirals, straighten up. In this new arrangement, the liquid crystal no longer rotates the polarisation of the light. So the screen goes dark. Apply a current, dark, switch off the current, light. If that were all there were to it, a screen would only be any good for producing Morse code, but in practice a display is divided up into many thousands of tiny segments, each of which has a kind of electrical grid to control it. Colour screens have each of these segments or pixels divided into three, corresponding to the primary colours. The grid enables a particular segment to be addressed and allows just its part of the crystal to have the rotation effect switched off and on – the result being that the screen can build up a complex picture depending on how each segment or pixel (a compact version of 'picture element') is switched.

This means that a screen can now display a complex image in colour. The computer I am writing this book on has a 2,550 × 1,440 display – 2,550 pixels wide by

1,440 deep – which means that the screen has a total of 3,672,000 of these segments, combining to make a very detailed image that is quite difficult to distinguish from a real view. Although there are now a whole host of competing technologies, each depending on a variant on the liquid crystal technology, they all work broadly on this principle, typically with a separate transistor controlling the current applied to each pixel to manage the way the display shows an image.

Viewing the fourth state of matter

An alternative to LCD that was more popular before it was possible to manufacture really large LCD displays is the plasma screen. These look like LCDs but are often significantly brighter. This brilliance comes at the cost of considerably higher power consumption and shorter lifetime than an LCD. A plasma screen is really like a huge matrix of tiny fluorescent light bulbs. Each little cell on the screen's surface contains a noble gas like neon and a small amount of mercury. The mercury is vaporised by an electrical charge, creating a plasma, a collection of ions – atoms with electrons missing or added. Unlike a gas, a plasma is very conductive. Electrons flow through the plasma and briefly excite the electrons attached to the mercury atoms which then drop back down, giving off ultraviolet light. When this high-energy form of light hits a phosphor at the front of the cell it makes the phosphor glow with visible light, creating the image.

Plasma displays are arguably in decline these days, but LCDs also see competition from LED screens. These have pixels made up of tiny light-emitting diodes (hence the name), each smaller than a pinhead. These diodes make use of a quantum effect in a semiconductor where electrons plunge into holes, giving off light as they do so. The most impressive thing about LED panels is that there is no limit to the size that can be constructed – they have been built 40 metres across for outdoor displays. For TVs and computer screens OLEDs (organic light-emitting diodes) are the most common form of LED, using an organic compound as the light-emitting layer. LEDs are increasingly popular because they can be used to produce thinner, lighter screens than LCDs with a higher contrast ratio than their older counterpart.

Quantum snaps

Another technology that has been totally transformed since the 1990s is the means of taking photographs. This change is reflected in the way the Eastman Kodak company, once one of the best-known brands in the world, was forced to go into protective bankruptcy in 2012 after the total collapse of the market for its films. The film cameras that were used all the way through the 20th century had changed only incrementally from Victorian photographic equipment, but quantum theory turned the technology on its head by capturing an image using electronics.

The digital camera has revolutionised photography, transforming its business model because there is no longer a consumable involved in taking pictures. It is ironic that the digital camera was actually invented at Kodak in 1975, but the company initially suppressed the product, realising that it would have a negative impact on its core film business. This was as short-sighted as the reaction of Victorian gas lighting companies which, faced with electric lighting, decided that all that was necessary to see off the competitor was to develop a better gas mantle. Kodak has paid a heavy price for its conservative stance. As a result of the move to digital we now take vastly more photographs than we used to – and of course, thanks to the incorporation of cameras in the ubiquitous mobile phone, many of us carry a camera at all times.

Light entering a digital camera is typically passed through a mosaic of tiny coloured filters, because the sensors used to detect the light don't distinguish colours. The sensors themselves work in two possible mechanisms. The earliest technology, used from that first camera in 1975 and still very common, is the charge coupled device (CCD). This is effectively made up of an array of tiny containers, each of which can hold electrons. As photons hit a region of the device, knocking electrons free, the number of electrons inside the cell builds, so there will be a higher electrical charge in a cell if it has been exposed to more light. When the image has been captured, the voltage of each cell is measured to produce the data that will be turned into a picture.

The alternative approach to CCDs is a 'complementary metal oxide semiconductor' (CMOS) sensor. In effect, the CMOS sensor is an integrated circuit with an array of light-sensitive diodes and amplifiers which react directly to the incoming light, rather than building an image over time. CMOS sensors have tended to take over the market for most ordinary cameras, as they operate faster and are cheaper to make than CCDs. However, CCDs are still used in some applications like high-quality video cameras, as they capture a whole image, where CMOS sensors usually capture a row at a time, which can result in odd visual effects when capturing high-speed moving images.

The ubiquitous interaction

These quantum technologies are now all around us. We take them pretty much for granted. There can be few houses indeed without a TV or a radio, a computer or a phone. If we take a trip to the laundry room, even the humble washing machine has computing power and a screen. Not only was Faraday right that one day the government would tax his invention, but those first steps in generating and using electricity were the opening steps in the transformation of our day-to-day lives.

Electricity, of course, existed long before we developed technology based on it. All living things have an electrical component to their internal operation. There is also electricity on the loose in nature, most dramatically in lightning. But there is another quantum effect that is much more obvious in the natural world and in some ways is even more remarkable than the electrical effect. This quantum marvel is responsible for the way we see, and why the Earth is habitable thanks to the Sun. It is even at the heart of the attraction and repulsion between electrically charged particles that lies behind the structure of solid matter, all mechanics and the way that we can sit on a chair rather than fall through it.

It is the interaction between light and matter.

CHAPTER 4

QED

Practically everything we experience is the result of light, matter or both. Light is much more than a phenomenon that enables us to see. It is light from the Sun crossing the vacuum of space to the Earth that carries the energy that keeps our planet warm enough to be habitable. Different frequencies of light perform the cooking in microwave ovens, support our radio, TV and mobile phone communications, and enable medics to produce X-rays and CT scans. And at a fundamental quantum level, photons of light are the carriers of the electromagnetic force that is responsible for the majority of the direct physical experiences we have, from being able to touch something to not disappearing through the floor.

Fire from the eyes

Light has fascinated human beings for as long as we have any record of what we thought and puzzled about – and no doubt was a mystery and a wonder long before then. Inevitably the first associations we made with light were in terms of sight. The prevailing idea in Ancient Greek times was that sight was made possible by a special fire in the head that projected a beam from the eye to the object being seen. (Built-in water chambers were thought to prevent this fire from burning us.) This might seem a crazy concept, because it implies that there is no need for an external light source to be able to see something – yet the Greeks rationalised this problem by saying that the Sun had a role in sight, but only in facilitating the beams from the eye.

If this now appears clumsy and in need of the application of Occam's razor, it came about by an approach that put philosophical structures above what was observed. Because sight was understood as something that we do to the world around us, it seemed inconceivable to the Greeks that it could be driven purely by an external source. They weren't willing to accept that we are just passive receptors of light that is already out there. A good example of the very different understanding they had of light comes from the Greek mathematician Euclid. He pointed out that if we are a looking for a needle on the ground, the light from the Sun is always falling on it. But we don't necessarily see it. It is only when the viewing 'light' from the eyes falls on it that it comes into view. The Sun facilitated, they believed, but the eyes' beams produce sight.

By the Middle Ages, Arab and European scholars alike had set aside this tortuous Greek reasoning and saw light as some kind of flow that came from a source like the Sun, reflected off an object as a jet of water sprays off a wall, and that came from that object to our eyes, enabling us to see it. They realised that the Moon did not shine with its own light but simply reflected the dominant light of the Sun. As the picture of light as something that flowed from place to place in straight lines become more sophisticated, it was expanded by the introduction of technology that helped us to manipulate that flow. The earliest optical technology had been mirrors, polished metal or stone that reflected light in a more concentrated fashion than an ordinary object, but it was the development of the lens that turned light into a more profound vehicle for exploring the universe.

The word 'lens' is derived from the Latin name for the lentil, reflecting the similar shape of a convex lens to the edible pulse. If light was something that travelled in straight lines (without worrying too much what that 'something' was), lenses could be used to bend that flow, to manipulate light, concentrating and magnifying the outcome. Soon the lens was making it possible to examine very small items that were not discernible to unaided eyesight, or to take in distant views, not only on the Earth but out in the heavens. Light was becoming a more versatile phenomenon, harnessed to human needs – yet very little progress was made in understanding what light was.

Mechanical light

French philosopher René Descartes, in 1664, was one of the first to give a rational scientific explanation for the way light moved, though his concept was easy enough to counter. He believed that all of space was filled with an intangible substance he called the plenum. (This might seem strange enough to call his theory into question, although time and again scientists have come back to a similar concept, whether it was the luminiferous ether, which we will soon see being used to explain light's motion, or the modern concept of quantum field theory, which is one very significant way currently used to model the nature of reality.)

Descartes imagined that when something gave off light

it set up a kind of pressure (as he called it, a 'tendency to motion') in the plenum. So, for instance, when we look up at the night sky and see a star, it was, Descartes thought, as if there were an immensely long rod connecting the star to our eye. The star presses on one end of the rod; the other end of the rod presses on the eye, producing the effect of vision. This model implied that light travelled at an infinite speed, which has been a subject of much debate for centuries. Descartes knew that it was certainly very fast – Galileo had attempted to measure the speed by sending a servant to a distant hill with a lantern at night, trying to time the light's two-way journey when he signalled to the servant and the servant signalled back. He found exactly the same timing was recorded when they performed the experiment standing next to each other. All the delay they were able to spot was caused by their response time. But whether light was instantaneous or just speedy was beyond the experiments of the day.

It was Isaac Newton, born nearly 50 years after Descartes, who made some striking discoveries about light, and replaced Descartes' theory with a more practical-sounding one – though Newton's own ideas were disputed and would soon be dismissed for over 200 years before they were found to be closer to the truth than anyone could have imagined. Newton thought that light was a stream of particles he called 'corpuscles'. These meant that light could travel through space without any invisible plenum or ether, making Newton's model pleasingly simple compared with most of the opposing theories; and according to him, light would be expected to travel at a finite, if extremely quick speed.

Unweaving the rainbow

Newton did not attempt to measure this speed, but he did perform a number of experiments that opened up our understanding of light and colour. When he was at Cambridge in 1664, aged 22, Newton attended a fair and bought a toy that would prove highly educational. At the time the university's private police force, the proctors, attempted to keep university members under control (and away from the city's taverns), which meant that the Stourbridge Fair, just outside the bounds of the proctors' remit, was an annual opportunity for the academics to let their hair down and have a little fun. It was there, from a stall selling toys and trinkets, that Newton bought a prism, a block of glass with a triangular cross-section.

The reason the prism was sold as a toy is that it had long been known that it would produce a pretty rainbow pattern when light shone through it. Newton was deter-mined to discover what was going on. The most popular theory of the time was that imperfections in the glass coloured the light as it passed through – something that seemed quite likely, as the quality of glass at the time was fairly poor (so poor, in fact, that when the German writer and science enthusiast Goethe tried to reproduce Newton's experiments he couldn't even see a rainbow). After playing with the prism for a while, Newton was inspired to obtain a second one to test out this theory.

He argued that if the spectrum was produced by imper-fections in the glass, then if he picked out a particular colour and sent that through his second prism, the colour would be modified once more. It wasn't – it stayed the same. Even more impressively, he consistently observed that the different colours were bent by different amounts by the prism. (There is some doubt about how much Newton achieved with the very poor-quality prisms he was working with, and how much it was a case that he was convinced what the result *should* be and reported it as such. The fact remains, though, that he got it right.) The observations he made inspired Newton to explain that white light was made up of all the colours of the spectrum, which were split out to different degrees by the effect of passing through the prism.

Once he understood the way that the colours were always present in sunlight, it opened Newton's eyes to the explanation of why we see an object as being a particular colour. When, for instance, white light from the Sun hits a red post box, the paint tends to absorb the colours in the beam, but it doesn't hang on to the red light. So when the light reflects back to our eye, it is only the red component that is left and the box appears red. Although there was some resistance to Newton's ideas, particularly from his arch-rival, Robert Hooke, Newton's viewpoint soon triumphed. But Newton had less success in persuading others of his notion that light was made up of corpuscles. Unlike the rainbow from the prism, he had no clear experiment to support his notion, merely a significant distaste for the alternative theory, and to be fair, there was a good reason for his suspicion.

Ripples in the ether

With Descartes' theory out of the way, the alternative to light being a particle was that it was a wave, like the ripples that spread out on a smooth pond when a stone is dropped into it, though operating in three dimensions rather than two. This was the opinion, for instance, of the Dutch scientist Christiaan Huygens. It was already widely accepted that sound travelled as a wave through the air, and the (admittedly limited) similarities between sound and light, particularly in their anthropocentric linkage through the senses of hearing and vision, helped support the idea that light also travelled this way.

Newton, however, quite reasonably struggled with the concept of the wave, because of the extra requirement it placed on nature. Particles can happily pass through empty space. But waves need a medium. A wave is fundamentally just a movement within a substance. Something has to do the waving. For sound it was obvious enough that the medium was air, and before long this was made very clear when a ringing bell was put in a jar and the air was sucked out. The bell could no longer be heard, because there was no air for the sound to travel through. But the bell did not disappear – it could still be seen. So the light had no need for air to create its waves.

Huygens imagined something like Descartes' plenum, but instead of being rigid, he thought that it was composed of lots of little compressible chunks, rather like space being full of tiny rubber balls. Light would pass through it as a series of little wavelets, moving from ball to ball.

Initially there was little experimental evidence to support either theory, with much argument over exactly what was happening in refraction, when light bent as it moved from one substance to another. But at the very start of the 19th century the apparent death blow for Newton's idea was Young's experiment with two slits that we have already met (see page 9). If light was corpuscles, as Newton said, Young expected to see two bright bars on the screen, one for each slit. But instead what was seen was a series of dark and light fringes. This just didn't make sense for particles. However, if light were a wave, as we have seen, interference between the two waves should cause the fringes that are observed.

Which way to wave?

Young also guessed (correctly) that the frequency of the wave – the time it took to get from peak to peak – was different depending on the colour of the light. In fact, that this was why there *were* different colours of light – and this was also reflected in his experiment, because changing the colour shifted the pattern on the screen, as would be expected if the wavelength was changing. But Young also made a suggestion that many took to be a step too far. It had generally been assumed that light had to be a wave, like sound, that made its oscillations back and forth in the same direction as the wave travelled, rather than jerking from side to side, like a wave on the surface of a pond or in a rope. But Young thought there was some evidence that light *was* such a transverse wave that wiggled side to side compared with its direction of travel.

This presented a real problem for the theory, because transverse waves can exist only on the edge of something. On the edge, the wave can stick out of the material unopposed, but if it tried to travel through the centre of the material it would soon be damped down by collision with the stuff around it. Yet whatever it was that was waving to enable light to travel, known by now as the luminiferous ether, the light travelled comfortably through the middle of it. How could the wave ripple from side to side? Although no one could explain how this was sustainable, Young had heard about the effect called polarisation where you could have apparently different 'types' of light associated with a direction at right angles to the beam's travel. This, Young thought, could only sensibly be explained if light was a side-to-side wave. Worrying about how this was possible would have to wait for a better theory.

More and more experiments confirmed that light did travel as a wave, though uncomfortably there was no other evidence for the existence of the ether needed for it to wave in. And the ether had to be a strange material indeed, filling all of space, totally undetectable to the touch, capable of vibrating and yet infinitely rigid so that there was no loss of energy as light waves passed through it.

Faraday's speculation

In 1846, Michael Faraday would give the first suggestion of why the ether simply wasn't needed. Although he seems to have been a reserved person in everyday social life, Faraday was a great science communicator and regularly lectured at the Royal Institution. According to legend, fellow physicist Charles Wheatstone was due to give a Friday evening discourse at the Institution. These stiff affairs were decidedly intimidating, especially as, by tradition, the lecturer was expected to rush out onto the platform and begin his talk with no introduction.

On the fateful evening of 10 April 1846, Wheatstone is said to have lost his nerve at the prospect of addressing such an audience and rushed off, leaving Faraday, the man responsible for the event, to fill in and extemporise. It's a good story, but probably not true. The reason this is thought to be legend is that Faraday had a week's notice to fill in for James Napier, who was the person actually due to speak. Faraday did give a brief paper on Wheatstone's invention of an electric clock, but then went on to speak on his own thoughts about light.

As we have seen, Faraday had come up with the idea of a field, extending out from a conductor, and producing, for instance, electricity as magnetic field lines were cut. Now, he suggested that light was a vibration – a wave – in those lines of force. Not a mechanical vibration in matter, but a wave in an insubstantial force field. As he put it in his lecture, his theory 'endeavours to dismiss the aether, but not the vibrations'. If the lines of force were more than imaginary aids but a real field (whatever that was), then it could carry vibrations without the need for a magical ether.

The interplay of waves

Michael Faraday never pretended to be a mathematician, and his ideas were more a visual representation than a detailed theory. But when James Clerk Maxwell (see page 50) came up with a mathematical description of the linked effects of electricity and magnetism, he also provided the final piece in the jigsaw of understanding the nature of light, filling in the gaps of Faraday's hypothesis. Strangely, Maxwell never did dismiss the idea of the ether (showing that even the greatest scientist can be more than a little conservative in his or her ideas). In fact it was by considering the ether as a fluid that he came up with his show-stopping result.

Faraday had shown that moving magnetism produced electricity, and moving electricity produced magnetism. Maxwell realised that a magnetic wave moving at just the right speed would produce an electric wave that produced the magnetic wave and so on ... but it would continue to haul itself up by its own bootstraps only if it moved at the speed of light. 'This velocity is so nearly that of light', Maxwell commented, 'that it seems we have strong reason to believe that light itself (including radiant heat and other radiations if any) is an electromagnetic disturbance in the form of waves propagated through the electromagnetic field according to electromagnetic laws.'

At this stage in our understanding of light, it couldn't be further from quantum theory. Although, as we have already seen, it was the understanding that light had to come in little packets that would form the origin of quantum theory, Maxwell's electromagnetic waves in the ether were entirely smooth and continuous. No packets here; nothing strange. Maxwell's assumption that the ether still existed (ignoring Faraday's dismissal of the need for it) was still the predominant view towards the end of the 1880s, when American physicists Albert Michelson and Edward Morley set out to study the ether. They planned to use the Earth itself as part of their laboratory.

An altar to the ether

The ether was considered the universal constant, something that was fixed and that everything moved through, giving a frame of reference to measure everything against.

If the ether were not there, as Galileo had shown in the concept known as relativity long before Einstein came on the scene, all movement would have to be considered as relative to *something*, because there was no 'preferred frame', nothing that defined what being motionless truly meant. The ether provided that reference. As the Earth rushed through the ether on its journey around the Sun, it should result in an effective flow of the ether past the planet, which meant that light should travel at slightly different speeds depending on whether you measured it in the direction the Earth was moving or at right angles to its travel.

Michelson and Morley's experimental set-up looked more like something that belonged in a medieval cathedral than a highly technical experiment in a late 19th-century laboratory. On a brick base was a circular metal trough filled with the alchemists' favourite substance, mercury. A wooden structure floated on this liquid metal, on top of which was positioned a large stone slab. So carefully constructed was the equipment that once it was set in motion at a stately rate of turning once every six minutes, it would keep going for hours without further intervention.

On top of the stone platform was an optical arrangement that split a beam of light in half, sent the two portions off at right angles, then recombined them, where a microscope made it easy to watch the interference fringes produced when the waves of light interacted. If there was any difference in the speed of light in the two directions, it ought to result in the fringes shifting as the great slab ponderously rotated. The outcome was a huge anti-climax. Nothing happened. What had set out as an experiment to measure a property of the ether ended up in dismissing its existence altogether.

Some would remain uncomfortable that there seemed to be nothing for the wave of light to wave in, while others reverted to Faraday's position that there was no longer a need for an ether, because the wave was simply a displacement in the insubstantial electrical and magnetic fields. However, once Planck's development of the quantum and Einstein's theory linking it to the photoelectric effect had been established, there was even less to be worried about, as no one thought that particles needed an ether and it seemed that light was, at least in some mysterious way, as much a particle as it was a wave.

A mighty spectrum

By the time the quantum theory of light was starting to be accepted, 'light' had gone from being a term for the stuff we saw with to covering a huge spectrum of electromagnetic radiation, running all the way from very long-wavelength radio, through microwaves, infrared, visible, ultraviolet, X-rays and gamma rays. The bit we can see is just a very narrow slice sitting around the middle of the range. All those distinctions are man-made. There is no boundary between, say, radio and microwaves, or ultraviolet and X-rays where something drastically changes. There is no difference other than frequency or energy between the radar we use to keep aircraft flying safely and the deadly gamma rays emerging from a nuclear blast.

If we think of light as a wave, we simply have an electromagnetic wave of shorter and shorter wavelength, or higher and higher frequency, as we move up the spectrum.

With a quantum hat on, the picture is even simpler, going from low-energy photons to high-energy photons – and all the differing capabilities, like, for instance, the way X-rays pass through flesh and damage DNA, are nothing more than a reflection of those increasing energy levels.

The dance of light and matter

What was emerging with the development of quantum theory was an understanding of the way that light and matter interacted. The original quantum idea had come from the way hot matter emitted light, while Einstein's work had been based on light hitting metal and knocking out electrons. It might seem the interaction of light and matter is only a small part of the physical world, but it proved to be one of the most important. All light is produced by matter. It is the interaction of light and matter that heats our planet and enables us to see. And it would be discovered that even as fundamental an interaction as the electromagnetic repulsion between your bottom and a chair that prevents you sinking through it is a result of a stream of never-observed photons passing between the atoms.

Our understanding of this interaction between light and matter, which was given the name QED, short for quantum electrodynamics, was the work of a good number of people, but one inevitably stands out. The groundwork came from the reclusive British physicist Paul Dirac, but the best-known approach that pulled together QED came from one of his contemporaries. He was the physicists' physicist, the man whom practically any working

physicist would put at the top of their 'most admired' list
– the American genius Richard Feynman.

The Feynman touch

Feynman's popularity combines a number of factors. He
was outgoing and a great communicator in a field where
many workers were (and still are) introverted – but he also
had a way of seeing problems in physics with a different
eye to many of his colleagues, not relying on existing ways
of doing things. This view seems to have been instilled in
the young Richard at an early age by his father, Melville.
Born in the dying months of the First World War, Richard
would be challenged by Melville to look beyond the labels
to what was actually happening.

Feynman studied at MIT before moving to Princeton.
While searching for a topic for his doctorate, he came
across a paper on quantum theory by Dirac. It set Feynman
thinking how to generalise what the British physicist had
done to produce a simpler description of quantum events.
Feynman took a very visual approach to his mathematics, and made use of the idea of particle 'world lines'.
In Feynman diagrams these are plots of the position of
a particle against time, with time as one dimension and
position in space the other. Traditional world line diagrams always have time going up the page and position
horizontally, but Feynman diagrams can be drawn in
either orientation.

So with time going up the page, a stationary particle
would be described by a vertical line, while one moving
at a steady speed would be a diagonal straight line. In
the real world, of course, particles can move in three

dimensions of space, but that would make a Feynman diagram four-dimensional, which is a bit of a pain to draw. To keep things simple we consider just the one arbitrary spatial dimension, but acknowledge that the rest are there too.

Quantum theory made it clear that a particle didn't have a definite location with time or trajectory. So Feynman imagined drawing every possible world line that linked the particle's start and end points. If you could pull together all these lines, each with its own probability attached, you would have a complete description of how the particle behaved. There were infinitely many ways for a particle to get from A to B, but Feynman realised this didn't have to be a problem. After all, the whole point of integral calculus was to provide a way of producing a finite end product from adding together an infinite collection of small quantities. And anyway, many of the paths would in practice cancel each other out, or would be of such a low probability that they could be ignored.

At the time Feynman's idea wasn't a practical tool for calculation, but it was enough for his thesis. He settled down to head towards a comfortable conclusion to his PhD when the Second World War intervened. In December 1941, the same month as the Japanese attack on Pearl Harbor, Feynman was seconded to work on a top-secret project, to produce a bomb based on a nuclear chain reaction. The atomic bomb.

Quantum destruction

While working with a team looking at ways to separate the uranium isotope U-235 from the much more common and chemically indistinguishable U-238, Feynman completed his doctorate, in part because he was required to complete this before he got married. He had been engaged to his fiancé Arline for some time, but the urgency was a result of her advanced tuberculosis. They were married in hospital in July 1942.

Meanwhile, Feynman's work on uranium separation was made redundant by a better method, and he was moved to Los Alamos in New Mexico to become a more direct part of the Manhattan Project. He agreed to go only if Arline could be found a hospital nearby. In the end, the closest they could achieve was Albuquerque, 60 miles away. As Feynman worked on the complex calculations required to specify the detailed construction of the bombs, Arline got weaker, dying in June 1945. Just one month later, with Feynman present, the first nuclear device was exploded in the Trinity test at Alamogordo, 200 miles south of Los Alamos.

Feynman made a huge contribution at Los Alamos, both in the physics and in keeping morale high among the scientists, who inevitably clashed culturally with the military. One of his ways to do this was to constantly challenge the petty regulations that beset military life, getting a reputation for being able to break in and out of the base unobserved and of being able to access locked filing cabinets and safes. But it was once he got back to civilian life that he was able to pick up the work that went into his doctorate and carry it forward to develop the ultimate theory of the interaction of light and matter, quantum electrodynamics – QED.

Quod erat demonstrandum

The theory was developed in different but parallel ways by Feynman, Julian Schwinger at Harvard, and, independently, by Sin'Itiro Tomonaga in Japan – yet it was Feynman's version that made the theory approachable as a way of understanding the interaction of photons of light and electrons. In essence practically every interaction of light and matter comes down to a combination of very simple elements. Either an electron loses energy and gives off a photon, or an electron absorbs energy and a photon disappears. Photons are born as electrons drop down in energy and die as they jump up in energy.

Although this seems a small part of reality, in fact QED covers a remarkably large swathe of our everyday experience, and does so stunningly well. Of all the physical theories we have, QED's predictions are by far the closest to what is actually observed. As Feynman delighted in pointing out, QED so well matches observation, it is as if we had a theory that predicted the distance from New York to Los Angeles to the accuracy of the width of a human hair. This apparent perfection, he noted, leaves us with a problem, because QED does not describe the common-sense world we think we experience. Instead it piles on the quantum weirdness with particles acting as if they considered every path possible at the same time and even sometimes acting as if they travelled backwards in time. And yet the theory is remarkably precise in predicting the actual outcomes we observe.

Feynman would use his world line diagrams, throwing in, where necessary, an additional complexity of an arrow attached to a particle like the second hand of a watch. This arrow rotated steadily with time, while the size of the arrow (or more accurately its square) indicated the probability of finding the particle at a particular location. This compass-needle arrow represented a property of a particle called its phase and provided the interface between a particle and a wave. The phase represented the position in the wave motion at that point in time and space. Its regular rotation with time enabled a particle to produce the observed wave-like effects.

Once he had photons on his diagrams equipped with their phase arrows, Feynman realised that every problem classical physics had thrown at the idea that light was a stream of particles – the way, for instance, it reflected off a surface or interfered with a second beam of light – was a necessary outcome of this structure. Once your particles had phase it wasn't necessary to think about traditional waves. This doesn't mean physicists *don't* still talk about waves. They are sometimes the easiest way to describe what is happening or to calculate the outcome. But they were no longer essential. It's important to bear in mind at all times that physics isn't about describing reality. It is about providing a model that predicts outcomes that match what is observed as closely as possible.

Light is ... light

I remember early on in my exploration in physics asking a professor: 'Yes, but what do we think light really is? Is it a wave, or is it a particle?' He groaned. Sometimes, Feynman seemed to come down firmly on the side of particles. He wrote: 'I want to emphasize that light comes in this form – particles. It is very important to know that light behaves like particles, especially for those of you who have gone to school, where you were probably told something about light behaving like waves. I'm telling you the way it *does* behave – like particles.' Feynman was right in the context of what he was saying. His approach to QED worked by considering light as a particle, and it worked well. Waves aren't necessary. But despite his dramatic wording he wasn't saying light literally was a particle.

Apart from anything, QED is a quantum field theory, treating light in some respects as neither a particle nor a wave but a field with a set of values across space and time. In fact that is how most working physicists will see light in practice. But even though some physicists would probably argue to the contrary, the idea of light as disturbance in a field is not a true description of reality either.

Light isn't a particle and it isn't a wave and it isn't a disturbance in a field. It's light. It operates at a quantum level that we can never directly observe or describe. Light bouncing off a mirror isn't like a tennis ball hitting a wall or like a wave hitting a blockage. Those are large-scale items that we can use to give a mental picture that represents what is going on, but they aren't what light is really like. And light isn't the outcome of a disturbance in a field – that is just a mathematical approach that happens to produce reliable results. All three are just representations – models, as scientists call them – that enable us to make predictions. Sometimes the wave model is easier to use, sometimes the particle model. From a mathematical viewpoint the field approach is most universal, but is often difficult to visualise. Each is sometimes helpful. None is a true picture of reality.

On reflection

To get a proper understanding of the revolutionary nature of Feynman's approach, and how the behaviour of these special particles can not only deliver the same results as considering light a wave, but can also better predict what really happens in some circumstances, let's look at a very simple optical set-up, one that was familiar as far back as the Middle Ages – a beam of light reflecting off a mirror. At school you were probably told the light beam or 'ray' travels in towards the mirror on a straight line, bounces off the mirror and heads out in a straight line at the same (but opposite) angle at which it arrived. This is a useful simplification, but it is by no means our best model of what is happening.

To begin with, light doesn't bounce off the mirror like a ball bouncing off a wall. An incoming photon is absorbed by an electron in the mirror, and then a second photon is emitted by the electron. It isn't the same light that leaves the mirror as originally headed towards it. But more significantly, the photon can take any route it likes to get from A to B. For instance, instead of bouncing off the middle of the mirror at a symmetrical angle it can head in at a much shallower angle, reaching much further across the mirror before it reflects, then shoot off at a much sharper angle. Every path it can take in this fashion has pretty well exactly the same probability of occurring.

So why do we see a beam of light reflect from the middle with equal angles?

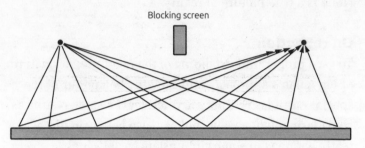

Fig. 3. Many paths reflection.

As it happens, the outcome is the same with the many paths and the traditional ray bouncing symmetrically. If you do the maths and add together all the possible paths, taking into account the compass arrows of phase, most of the paths will cancel each other out, resulting in the path that we expected in the first place – reflecting at equal angles. However, we can't throw away the more complex picture. If you take a mirror where light is reflecting from the middle and chop out the middle section you won't see a reflection. But put a series of dark lines on one side, leaving only the paths where the photon would have similar phases, and the reflection starts again – at an angle that seems totally crazy for reflection as we expect it to behave from our experience of bouncing balls.

A photon does, in a probabilistic sense, take every single path for the reflection; we just don't generally see the outcome because of phases cancelling.

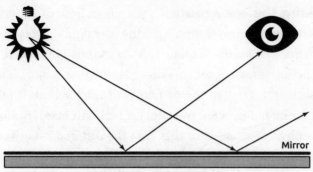

Normal reflection at equal angles

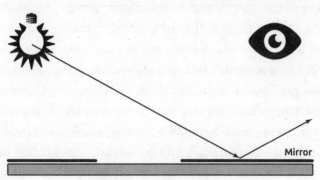

Nothing seen with centre missing

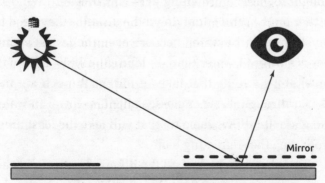

Reflection seen again at strange angle when section has dark strips

Fig. 4. Missing mirror reflection.

Taking the easy route

There's another revelation from the way light reflects from a mirror that seems to relate to a very fundamental aspect of the universe – it tells us that the universe is lazy. This is sometimes called the principle of least action. It tells us, for instance, what trajectory a ball will take. 'Action' in a physics sense is in this case the difference between the potential energy – the energy due to the ball being suspended in a gravity field that pulls it towards the Earth – and the kinetic energy of its motion. The path the ball takes is the one that minimises this 'action'. Similarly, light obeys the principle of least time, taking the route that will get it to its destination in the shortest time.

It might seem that this will always involve travelling in a straight line – and it often does. But where, for instance, light moves from air to glass or air to water, it bends in the process called refraction. In this particular circumstance the principle of least time is sometimes called the *Baywatch* Principle, as lifeguards understand the concept that seems to guide the path of light. Rather than run straight towards a drowning person, a lifeguard will run further on the sand to cut down the distance they need to travel through the water, because even the fastest swimmer is slower in water than on land. Similarly, light will travel along a route that takes it further through air and less far through water or glass, as light is slower in water and glass. It follows the path that will take the least time, producing that refracting bend.

Now when you look at all the different paths light can take getting from A to B while reflecting off a mirror, the paths that are near the centre of the mirror (numbered 5 to 9 in Fig. 5) are the ones taking the shortest time – and as you move away from the centre, to begin with, the path length increases only quite slowly. That means the photons following those central paths have phase arrows pointing in roughly the same direction, so they reinforce each other. As you get further away from the centre, the length of path increases more and more quickly, so there is more chance for the arrow to be facing in a very different direction and for the phases of a photon to cancel out. Richard Feynman came to his approach to QED from thinking about the principle of least time and its implications for photons.

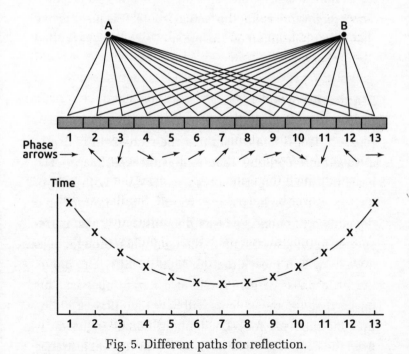

Fig. 5. Different paths for reflection.

This is also why we say that light travels in straight lines, and why we can usually get away with considering only straight lines from A to the mirror and mirror to B. In reality, there is a probability for a photon to take every single path between A and B, including going in the opposite direction, travelling to Paris, taking a trip round the Eiffel Tower and returning in a wiggly flight like a drunken fly. But these routes run counter to the principle of least time, and as soon as the photon wiggles significantly from the straight line it gives the phase clock a chance to get far enough away from that of a straight-line photon to have something to cancel it out – but all the very nearly straight-line routes have phase clocks pointing in almost exactly the same direction and add together. The straight line is not the 'actual' route, it is just the one that is left behind when the other clocks have cancelled each other out.

Magic mirrors

In the real world, reflection is often rather more complex. (In fact it is a truism that the real world is nearly always more complex than a physics experiment.) The basic reflection experiment is usually done with light of a single colour to keep things simple. The colour of light depends on the energy of its photons, but it also corresponds to the wavelength of the light when thought of as a wave, which means that for light of different colours the little phase arrows rotate at different speeds. This means that different colours will have their phases adding up constructively for reflection at different angles away from the centre. If we force the photons to go at a strange angle by selecting just some of the possible phases, by taking away the centre and using a series of dark lines known as a refraction grating (this is the same situation as Fig. 4 above, but with multiple colours), the different colours will split out. Shine white light on one of those special mirrors with the dark lines and you will see a rainbow.

This is an experiment anyone can do at home, because we have lots of discs with tiny slots in their surface that provide a similar effect to those dark lines – CDs and DVDs. Hold a disc at an angle in white light and you will see rainbow effects that are the result of exactly this kind of reflection in an unexpected direction predicted by QED.

Every which way

We can also use Feynman's approach to take a subtly different way of looking at the quantum version of the two-slit experiment, whether it involves photons or electrons. The way we looked at it in Chapter 2 was to say that a photon didn't have a location. All that existed before, for instance, it hit the screen and was registered, was a wave showing its probabilities, which is why in passing through the slits, the wave could cause interference. Feynman's way of looking at it would be to say that we can think of the photon taking every possible path between its source and the point on the screen where it is detected.

Think of that for a moment, and consider how mind-bogglingly outrageous this is. William of Ockham, who came up with the 'Occam's razor' idea we have already met ('Ockham' is spelled differently for historical reasons), would have a fit. His principle says that with nothing else to guide us, we should go for the simplest theory (to be precise, 'entities should not be multiplied unnecessarily'). Yet here in Feynman's imagination is an entity being multiplied an infinite set of times. We are not just saying that the photon takes every simple straight-line path, like passing through both of the slits in as close as possible to straight lines before ending up on the screen. We are saying, for instance, that the photon heads off into space, takes a trip around the Sun and comes back to land on the screen.

However, as we have seen, the vast majority of these paths will cancel out or have disappearingly low probabilities. Nonetheless, if we imagine the photon taking every path, then it can do everything the probability wave can, including producing an interference pattern. It's not strictly accurate to say that the photon 'does' take every path, as there are probabilities attached to the paths and the word 'does' seems to imply 100 per cent certainty, putting us in danger of the erroneous statement that the photon is in more than one place at the same time – but this is a problem that English has in being applied to quantum problems, rather than anything wrong with the picture. The outcome – and Feynman's method – is often described as 'sum over paths' (or for mathematicians, who get embarrassed by using such simple words, 'path integral formulation') because we literally consider the quantum particle to have taken every possible path and add together the results, bearing in mind the phases and probabilities.

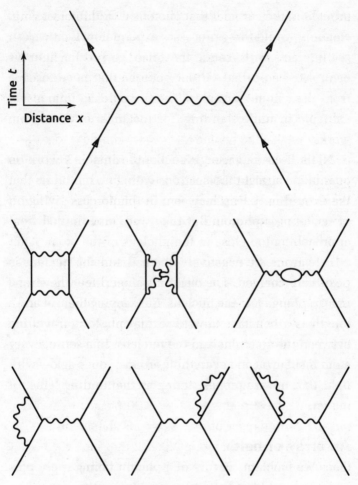

Fig. 6. A few Feynman diagrams
for simple two-electron interaction.

The glue of the universe

As he developed the diagrams that would be permanently linked to his name, and that are massively useful tools in quantum physics to this day, Richard Feynman brought in the particles that photons would interact with. Not only would these processes explain how light we can see interacts with matter, they also worked for light we can't see – so-called virtual photons that never escape from their home particles. These would be immensely valuable in understanding, for instance, how an atom works.

Niels Bohr had used Planck and Einstein's work on quanta to suggest that electrons were fixed in orbits that they could only leap between, in the process giving off or receiving a photon. But they were also limited from approaching too close to the nucleus of the atom. After all, electrons are negatively charged and the nucleus is positively charged. The electrons should feel the strong urge to plunge into the nucleus. But keeping them at arm's length is a constant flow of virtual photons travelling between the electrons and the nucleus. In a sense, every atom inside you and everything around you is aglow with light that never escapes, acting as the binding 'glue' of matter.

An array of fields

Once we have the picture of a photon taking every possible route, and of the little clock indicating the phase (direction of the hand) and probability (size of the hand, or more precisely the square of the size of the hand) at every point in the space the particle can be considered as passing through, we have returned in a way to Faraday's view – but with a mathematical reality he never envisaged, described by Schrödinger's equation and the other equations of quantum mechanics. Faraday thought of electricity and magnetism in terms of field, and our infinite array of clocks is nothing more or less than an alternative way of describing a field. QED is called a quantum field theory because it takes the 'field' approach to describing what is going on – and such theories are very common in modern physics, because they can often be the most practical mathematical way of explaining the outcomes we observe.

It's easy to think that fields don't really exist – after all, they are very strange things, mathematical entities that stretch throughout the universe having different values at different places. Only a mathematician or theoretical physicist could love them. They reek of overkill. They have none of the easy acceptability of the idea of a particle, with its apparent resemblance to a small ball travelling from A to B. But we have to bear in mind that quantum particles bear no resemblance whatsoever to a ball in the way they behave. Using a field is no more right or wrong than using a particle. They are both entirely equivalent models – but the field approach is often the most practical to calculate results.

The colour problem

Feynman diagrams are very useful to understand visually what is happening in an interaction between particles, but they also became the mainstay of calculations of the way particles would behave, assigning probabilities to each diagram to produce an overall outcome. In principle this is a lengthy task, as there is an infinite set of different interactions, but the probabilities drop off very quickly with the more obscure combinations, adding more and more virtual particles, which means that generally speaking it is possible to go to only a couple of extra layers of detail and ignore the rest. And Feynman diagrams have been used this way to the present day. But there are some issues that remain.

Specifically, problems arise when extending the scope of the diagrams beyond electrons and photons to include the interactions of nuclear particles. The relatively massive particles that form the central part of atoms, neutrons and protons, are made up of triplets of fundamental particles called quarks, which are linked by a flow of gluons, the quarks' equivalent of the role that photons play in electromagnetism. The parallel theory to QED for quarks and gluons is quantum chromodynamics, or QCD. The name comes from the fact that gluons come in three different 'flavours', distinguished by the colours red, green and blue. (Gluons aren't actually coloured – the concept is meaningless, as they don't interact with light – the colour names are purely arbitrary.)

This added complexity of colour compared with the colourless photon means that the Feynman diagrams for QCD become significantly more complicated, and the sheer number that have to be considered spirals out of control. Just dealing with an event where two gluons collide to produce four other (less energetic) gluons, a very common process in a particle collider like the Large Hadron Collider, requires 220 diagrams to be assessed before the contributions become negligible, resulting in a calculation with millions of terms. There had to be a better way to handle quarks and gluons, but one would not emerge until the 21st century.

A jewel of physics

A hint had come in the 1980s, when Stephen Parke and Tommy Taylor, working at the Fermi National Accelerator Laboratory in Illinois, managed to come up (using more than a little guesswork) with a simple mathematical expression that replaced all the diagrams and calculation terms for that two-gluon interaction. It took over twenty years to get beyond this first leap of inspiration, but in the mid-2000s, by using a mathematical tool called twistors developed by British physicist Roger Penrose, a way of mapping space and time into a different kind of space using complex numbers, an increasing number of short-cuts were developed – but at this stage no one knew why they worked.

Recently it has been discovered that these shortcuts were dependent on a concept with the imposing name of a positive Grassmannian. This was a bit of maths developed in the 19th century that is a multi-dimensional version of the inside of a triangle. In the triangle we are operating in two dimensions, and the space inside is a region bounded by intersecting lines. The more general Grassmannian is multi-dimensional, with the space demarked by the intersection of planes. And for particle interactions, a Grassmannian of the same number of dimensions as there are particles involved is required to describe the scattering process.

The final part of the journey to calculating the scattering amplitude (which describes the way the particles interact), traditionally the role of Feynman diagrams, is to create a new object, a so-called 'amplituhedron'. This is a structure that has been described as a 'jewel of physics' in that it brings together the Grassmannian structures to produce a single object whose structure has the solution coded into it, as its volume equals the scattering amplitude. The amplituhedron for an eight-gluons interaction, for instance, can be sketched with a handful of lines, but is the equivalent of 500 pages of calculations. It's early days – this theory was developed only in 2013 – but a single amplituhedron can correspond to hundreds of pages of calculations and may provide a way to reach values that simply weren't accessible using the conventional Feynman diagram calculations.

CHAPTER 5

Light and magic

Once we have the insight of QED, every interaction between light and matter becomes a quantum affair.

Let there be lights

Think of the humble light bulb. Thomas Edison and Joseph Swan, who shared the credits for inventing the light bulb, really weren't bothered exactly where the light was coming from, they just wanted a workable product to replace the dirty, dangerous gas jet lighting. Swan (who bore a startling resemblance to Brian Blessed), working in Newcastle upon Tyne, got there first by a few months in 1879, but the more astute Edison knew the power of patents, and as soon as his own patent was established he sued Swan for infringement.

All too often the history of invention is littered with cases where the winner of litigation was simply the one with the most money, but this time the courts recognised the validity of Swan's claim of precedence and found against Edison. As part of the judgement, Edison had to concede that Swan independently invented a working bulb ahead of his rival, and grudgingly set up the Edison and Swan United Electric Light Company. It's not that Edison's contribution wasn't significant, both in independently devising a bulb based on a carbon filament and keeping it working for considerably longer than Swan did, but it does seem a little unfair that most people would name the inventor of the electric light bulb as Edison.

Whatever their claims – and not to mention the long-drawn-out battle Edison would have with Westinghouse, pushing the benefits of his direct current electrical supply over Westinghouse's alternating current system devised by the great Nikola Tesla – all that Edison was really interested in was keeping a hot filament glowing without it burning away. Yet the process behind that invention was just as much a quantum one – the emission of photons as electrons that had absorbed energy dropped back to a lower level – as any sophisticated electronic device.

The window puzzle

There was, at the time Edison and Swan were working, no real puzzle to incandescent light, but once it was known that light came in the form of photons, QED proved necessary to explain an effect that had baffled Isaac Newton: partial reflection. The simplest example has to be a glass window. A piece of glass lets through some of the light that hits it, but reflects some of it back. If you stand in front of a glass shop window you can usually see both what's on the other side and yourself reflected in the window. Light both passes through the glass and reflects off it. Specifically, light from you is reflecting off it, so you can see your reflection. But people inside the shop can also see you through the glass – so some of the light from you is passing through.

Because Newton thought of light as being made up of particles, or corpuscles as he called them, this proved a real problem for him. If light was waves, you could imagine a part of the wave going through the glass and a part being reflected. But particles like photons don't split into two like this. Either a whole particle reflects or a whole particle passes through. And the question Newton couldn't answer was why a particular particle would 'decide' to pass through or reflect. What made the difference? Because clearly some did one thing and some did the other.

An obvious assumption might be that it has something to do with the surface of the glass – perhaps that it was rougher and more scratched in some places than others. The glass of Newton's day was, after all, anything but uniform. But Newton could dismiss this with the throwaway line that it couldn't be true 'Because I can polish glass'. Newton spent a fair amount of time making optics, which meant polishing the surface of lenses and mirrors. As he polished up the surface he made finer and finer scratches, changing the surface and reducing the possibility of reflections being caused by unevenness in the surface. And yet the glass happily continued to reflect.

It is handy that this partial reflection process does work. Technically, a device that splits a beam of photons in this way is called a beam splitter, and while the effect might be irritating sometimes (for instance, when the reflection is off your laptop screen or a car windscreen), it is also very widely used in optical experiments, in head-up displays, in devices that make very precise measurements and some forms of camera, and even in special reflectors in spotlights, used to reduce the amount of infrared in the spotlight beam, which can cause the equipment to overheat.

This beam splitting is a purely quantum effect. There was no point in Newton ever stretching his brain cells in an attempt to find a rule or an explanation for why particular corpuscles (or photons as we would say) reflect while others pass through, because there *is* no reason behind the selection, other than the probabilistic nature of quantum theory. We can say that a photon has, for instance, a 10 per cent chance of reflecting off a particular surface, but we can never know what a particular photon will do until the outcome has occurred.

Thickness matters

Something even stranger (unless you take a quantum viewpoint) is that the reflection from the front of a piece of glass depends on how thick the glass is. Change the thickness while leaving everything else the same and the percentage of photons reflecting will alter. You can imagine some kind of wave-based explanation for this, where the wave is already passing through the whole sheet of glass – but it is very hard to understand if you think of photons as little point particles. How can they possibly know the distance to the back of the sheet of glass when they 'decide' whether or not to reflect?

Luckily for us, we know more about quantum particles, and specifically that their probability wave spreads out over time. Just as the probability wave of a single photon can be influenced by the presence of both slits in the Young's slits experiment, so the probability wave passes through the whole sheet of glass. This means there is a probability for the photon to reflect from the back of the glass, and to come back to the front. At this point, just as with Young's slits, the phases of a photon reflecting off the front and reflecting off the back will combine, depending on how much their 'clocks' have had a chance to rotate. There is only one photon, but as usual it doesn't have a location and all we can work on is the various probabilities. By combining the squares of the size of the arrows, we reach the probability of the actual reflection happening.

As the thickness of the glass increases, the amount of reflection increases up to a maximum, then decreases all the way down to practically nothing as we reach a position where the two phase arrows pretty well cancel each other out. As we get thicker glass still, the percentage reflecting will increase again, up to the maximum, then reduce, and so on. In reality what is happening is that the photon has a chance of interacting with every electron in the body of the glass, which will absorb it and then re-emit it in a new direction, a process called scattering. But as it happens the various probabilities and phase arrows add up and cancel out in such a way that the effect is as if the photon was reflecting from either the front or back of the glass.

This effect has a decorative application in the real world. Whenever we see something iridescent – think, for instance, of the rainbow colours you see in a thin film of oil on the ground – what is happening is that, because the photons of different-coloured light have clocks that rotate at different speeds, the probability of reflection for a particular thickness of oil is different for different colours. As the thickness varies slightly across the film, you see different colours reflected. When you see a piece of iridescent jewellery or pottery glaze, it is making use of this dramatic quantum effect.

The quantum cheat

Another example of QED in action is the lens, a piece of technology we use every day in spectacles and telescopes, in cameras and in the lens that is built into our eyes. What the lens does is cheat on the principle of least time. Imagine we look at two of the infinity of routes a photon can take going from something you are looking at – this page, for instance – to the retina of your eye. One possible route is a straight line from the object to the retina. Another possibility is for the photon to head upwards away from that straight line, then suddenly change direction and head back to hit the same point on your retina.

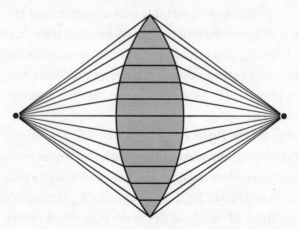

Fig. 7. Lens: shorter paths spend more time in glass.

From the principle of least time we know that the photon has a much higher probability of being found on the straight line, because it is the quickest route. If it took the route heading up a fair way then changing direction, it would be easily cancelled out by the phase arrow of another clock. But now let's put a bit of magic material in mid-air that slows photons down. We will make it thickest at the centre, where the straight line passed through, and thinner at the point our other route changed direction. So we are adding more time to the straight route than to the longer route.

The outcome of adding this magic material is that the straight-line path now takes the same time as the diverted path. Both routes now take the least time, and because they take around the same time, the phase arrows of their photons will add up. The photon will have a significant probability of taking both routes and arriving at the same point, so we have increased the chances of light turning up. We've focused the light. Using just the principle of least time and our little phase arrows, we have invented the lens from scratch. And any lens – including the lens in your eye – is shown to be a powerful quantum device.

Powered by light

It would be impossible to number every quantum device we use that depends on QED. Clearly this applies to every sort of lighting, but think also for instance of the solar panel. As we saw in Chapter 2, photosynthesis is a process depending on a quantum interaction of light and matter, and a form of quantum catalysis, but there is more conscious use of quantum effects in the technology we use to generate electricity from light.

As we have seen, the photoelectric effect, where a photon blasts an electron in a metal up to the conduction layer, was one of the key discoveries to inspire quantum theory. To be practical, though, the material used in solar cells (also called photovoltaic or PV cells) is a semiconductor like silicon, where the effect of pushing the electron up is to produce an electron/hole pair. The cells are structured like a diode, so the current can flow in only one direction, and joined together in a large matrix to produce sufficient electrical energy to be worthwhile.

There are a whole range of photovoltaic cell technologies, from the traditional, chunky silicon-based solar cell panels to cells that are printed on a film. The printed cells are much cheaper than a traditional structure, but also a lot less efficient – they turn less of the incoming light energy into electricity. A typical solar cell has around 25 per cent efficiency, with the best (but not yet commercially practical) current cells getting close to 50 per cent. By contrast, the film cells can be as low as 1 to 2 per cent efficiency – but can be produced for a fraction of the cost.

There was something of a media stir in 2013 when it was announced that solar cells produce more electricity when music is played to them. Entertainingly, it turned out that the cells preferred pop music to classical. This was specifically a variant of the film style of cell based on zinc oxide. These cells usually manage only around 1.2 per cent, but by building in little oxide 'hairs' that vibrate with sound, the efficiency can be nearly doubled. There's more energy produced from high frequencies than low, so pop music with its greater preponderance of high frequencies wins the race. However, there is no danger of solar farms with speakers blasting out the latest chart sounds to encourage generation. Significantly more energy is wasted pumping out the music than ever primes the cells to greater activity. This technology would, however, be useful in a naturally noisy environment – near an airport, for instance.

The moving quantum writes ...

One other example to show the breadth of QED in action. However you are reading this book, you will be making use of the quantum interaction of light and matter. It could be that you have a paper version, the oldest of our portable technologies for making a large amount of information accessible. Writing in all its forms is arguably the most transformational technology we have ever developed. And even though we receive a lot of information these days by other routes – for instance, watching videos – there is still a massive amount of information that reaches us in written form, whether on our electronic devices or through more traditional means.

As a form of communication, writing forms part of an activity that is common across living things – but it is a very special way of going about it, because writing takes away the limitations of space and time that restrict natural communication. I have books on my shelf (often via intermediary translators) written by the prehistoric Israelites responsible for the Old Testament of the Bible, Ancient Greek philosophers like Archimedes and Aristotle, the medieval scribes of the *Anglo-Saxon Chronicle*, Galileo, Newton and, yes, Brian Clegg. There are probably more books on my shelves bringing me communication from dead people than living – and certainly very few of them, with the exception of Terry Pratchett (and Brian Clegg), live near to my home.

Computers and the internet stretch this lack of locality even further. Emails can arrive from any place on the globe. The internet allows me to burrow into information from many places and times. And writing makes possible far more than the irritating spam email that fills my inbox. Without written documents we could not have developed the structures of science – our only means of understanding the universe would be orally transmitted myth. With no way of transmitting ideas across distance and centuries we would be constantly re-inventing the wheel (literally and metaphorically). And all this is based on the QED technology of writing.

From star to brain

On a traditional printed page, getting the information from the paper to the brain involves what seems a simple enough process, but nonetheless involves a powerful quantum chain of events. Let's imagine you are reading this book by daylight. Millions of years ago, high-energy photons were released from nuclear reactions deep in the Sun. Over the years they have been working their way outwards through the star's dense structure, being constantly absorbed and new photons emitted. Eventually they reached the Sun's surface, which is heated to around 5,500°C, making it glow white, spraying out many billions of photons every second.

Those photons cross space largely untroubled before they reach the Earth's atmosphere. Here they will be scattered by the gas molecules in the air (primarily nitrogen and oxygen), absorbed and re-emitted in new directions. High-energy photons are more likely to be scattered, so the unscattered photons we see in the Sun's disc appear a lower-energy yellow, while the scattered blue photons give a colouration to the whole sky. A mix of direct and scattered photons arrive at the page of your book, producing a white light. The photons are absorbed by the atoms of the paper and ink, increasing the energy levels of the atoms by boosting electrons to a higher orbit. In the black ink, much of this energy will go to heating up the molecules, but for the white paper many more photons will be re-emitted towards your eyes.

There is more quantum processing at work as the photons' paths are shaped by the lens of the eye, absorbing and re-emitting photons according to the rules of QED. There will be absorption and re-emission as the photons pass through the jelly-like inner substance of the eye, until they reach the retina. Here the photons hit special cells that contain a collection of photoreceptor molecules, where the absorbed photons set a collection of electrons on their journey to the optic nerve and finally into the brain, where the mixed signals are transformed into a visual image and the strange markings on the paper can be interpreted as a means of communication.

The electronic word

So much for the paper book, but the chances are increasingly high that you will be reading this book on an e-reader. When I first started writing there was no such thing as an e-book, but now they account for over a third of all sales of my books and that proportion is likely to increase. Many of those e-readers are software running on a computer or tablet using a conventional LCD screen (see page 72). I usually read e-books on an iPad, for instance. But dedicated e-readers like a Kindle or a Nook instead make use of e-ink.

This is a passive visual technology. Where an LCD pumps out photons towards your eyes, an e-ink reader sits there like a paper page, waiting for photons from the Sun or another light source to carry the message towards your brain. Although e-ink displays are much slower to refresh than LCD and limited in their graphic capabilities, they have two big advantages: they don't use energy to keep the image in place, only to set it up; and they can be made much less reflective, ideal for use in natural light.

A typical e-ink display uses a thin layer containing oil in which float bright white titanium dioxide particles. The pixels that make up the display are controlled by a pair of electrodes, top and bottom, with the top electrode transparent. When the electrode is charged up to attract the white particles to the top, the pixel shows white, but when the particles are attracted to the bottom electrode they fall to the bottom. Depending on the specific technology, either the black pigment is present in the oil, or there is a set of black particles that have the opposite charge to the white and so now drift to the top. Each pixel has a corresponding transistor and capacitor on a thin film forming an 'active matrix', the same addressing technology used in most LCDs, which actively ensures that the pixel has the correct value.

A special ray

From the electric light bulb to electronic paper, the possibilities for quantum optical technologies are endless. And to date, these devices have relied on the simple interaction of photons and matter with the assumption that all photons are equal. But a discovery from the 1950s would make it possible to produce light that made use of those photons in a very special way. Just as electronics made use of quantum theory to transform the way we control electricity, this new technology would use explicit knowledge of the quantum nature of photons to produce a kind of light that had never been seen before.

A light that could be used to decode music, to perform medical operations, or even to kill.

CHAPTER 6

Super beams

In the cold winter months of early 1973, two seventeen-year-olds quietly took over an unused storeroom in their school. Not untypically for teenage boys, they put up signs on the door proclaiming, 'Do not enter! Danger of death! Extremely high voltages!' – but unlike the warning signs that adorn the bedroom doors of many teenagers, these were more than for show. The pair really were dealing with technology that had the potential to kill.

Light school

I was one of those seventeen-year-olds. My school then encouraged would-be Oxbridge candidates to take their A-levels a year early, leaving us free in our final year at school to concentrate on the entrance exams. But these were taken at the end of the first term. If the school could encourage us to stay on for the rest of the academic year, they got significantly more money from the government. So they tempted us to remain by putting on all sorts of exotic classes – and by allowing science students to take on a challenging project. A friend, David Ball, and I decided that we were going to build a laser.

This was, at first sight, decidedly challenging. It was, after all, little more than a decade after the world's first laser was built. But we were following detailed instructions published in a 1970 *Scientific American* that guided the reader through the process of building a dye laser.

We didn't make it. In the end, we ran out of time. Our device never worked. But we got some excellent experience of building scientific equipment. We constructed the dye chamber, with its end mirrors, one half-silvered, suspended from finely adjustable metal mounts. We built a high-powered flash tube that zapped the remaining gas in a thick transparent tube, partly evacuated by attaching it to an air pump, with the output of a pair of huge capacitors. We even got the mirrors aligned, one of the trickiest tasks in building a successful laser (though I have to confess we cheated by using an off-the-shelf laser to guide the alignment).

The flash tube we did get working – a real scientific device built from scratch. Our flashes were dramatic in the extreme, giving us the masterful feeling of having tamed indoor lightning, and the great grey capacitors, each the size of an oil can, borrowed from the University of Manchester Institute of Science and Technology, were the reason for the sign. The charge they generated would have killed instantly if you touched the terminals. (I can't imagine schools allowing two students to play with such dangerous technology unsupervised these days.) But whether we would ever have got the laser itself to function I don't know. Certainly it proved a non-trivial task for the early workers in the field, as becomes obvious once you explore the history of this most iconic quantum device.

The Russian amplifier

One of the points of origin of the tangled tale of the development of the laser came in Russia in October 1954. Alexander Prochorov and Nikolai Basov published a paper in the Russian language *Journal of Experimental and Theoretical Physics*, describing an as-yet theoretical process based on a prediction that Einstein had made in 1916, shortly before he became totally engrossed with general relativity. The Russians believed that using the process Einstein described, a suitable material would act as an amplifier for microwaves (lower-energy photons than visible light), multiplying up the weak stream of photons to make a more powerful beam. This process would become known as microwave amplification through the stimulated emission of radiation – making the theoretical device a 'maser' for short.

What Einstein had realised was that when hit by light of the right wavelength – which in Basov and Prochorov's experiment happened to be in the microwave region – an electron could absorb the energy of a photon and sit in a semi-stable state, like the cocked hammer of a gun. If a second photon happened to hit that same electron while it was still in this state, then the electron would emit two photons, amplifying the incoming light as the photon triggered the electron's double leap downwards. If this took place in a reflective chamber, where the photons continued to pass through the medium, hitting electrons repeatedly, the result would be to build up a whole collection of atoms with electrons in an energised state ready to drop back down together in a chain reaction producing a stream of photons, synchronised by the process so that their QED phase arrows were all pointing the same way and moving together, making them 'coherent'.

For Einstein this was little more than an interesting piece of theory, though in the 1930s a Russian scientist,

Valentin Fabrikant, took things further in his imagination. He wondered what would happen if you had a material where the majority of the atoms were in an excited state, an inversion of the usual situation. If you then sent photons of the same energy through the material, it should trigger a mass emission, amplifying the original light (whether visible or, perhaps more usefully at the time, radio or the microwaves of radar) into a much more powerful beam. It was this concept that Prochorov and Basov had described in their paper, and that Prochorov described in ammonia gas at a conference organised by the Faraday Society in Cambridge in 1955. It was a concept that left one member of the audience stunned.

That's my maser

His name was Charles Townes, and the shock came because he had already built an ammonia maser. Unaware of the publication in a journal still largely concealed from the West by the Iron Curtain, Townes thought that he was the first to work on masers – and he certainly was in terms of constructing the device. But he did not publish details of his work on the maser until August 1955, giving Prochorov and Basov priority on the theory. Townes had worked on microwaves for a Bell Labs radar project during the Second World War. The powers-that-be wanted more and more detailed radar readings, which meant using microwaves of a higher and higher frequency, all the way up to 24 gigahertz (waves that oscillated 24,000,000,000 times a second). Following up on studies showing that the gas ammonia was easily excited by light of around this frequency, Townes had hoped that he had the solution to producing such pinpoint radar, only to discover that water vapour in the air readily absorbed this frequency, making it useless for radar. There wasn't a lot of point in developing a radar system that was stopped by an unavoidable component of air.

After the war, Townes moved to a professorship at Columbia University and continued his studies of the interaction of these high-frequency microwaves with molecules like ammonia. Working with a postdoc called Arthur Schawlow, Townes had originally considered working with ammonia molecules that were excited by giving them extra rotational energy, which would have required a very high-frequency source, but settled instead on the more conventional excitation of vibration, for which those microwaves around 24 gigahertz would be ideal. (In all cases, excitation is fundamentally about pushing electrons up to higher energy levels, but when this happens in a molecule it can result in a quantised rotation or vibration because of the change in the energy of one of its component atoms.) By now working with Jim Gordon, come the spring of 1954, Townes had achieved a working model of what they named a maser, and was ready to tell the world about it by the time he had the shock discovery of Prochorov's work in Cambridge.

It would be appealing if the story of every scientific discovery were a nice, neat tale of an individual or team working against all odds and producing a result that no one else has thought of, but in reality there are usually many contributors to a development (in what I've described above I've already omitted several other minor breakthroughs along the way), and it isn't uncommon for two or more totally unconnected teams to hit on similar results at about the same time. In this particular case, the distinction is quite clear – Prochorov published first on the theory (and publication is required to be considered to have got there first), but Townes was first with a working device. Before long, though, there would be a messier dispute over the maser's far more significant successor, where even to this day the priorities are not considered cut and dried.

A maser for light

The maser was an impressive device, and even though the ammonia maser did not produce the wonder precision radar that Townes had hoped for, its unparalleled precision in the frequency of its radiation had immediate applications in improving the design of atomic clocks. But it was swiftly realised that the concept had the potential to be far more dramatic when it was applied to visible light. The race was on to design a visible maser – and race it was, as American physicists Art Schawlow and Gordon Gould would enter into a ferocious patent battle over who came up with the concept first.

Schawlow and Gould were by no means the only ones to work towards the dream goal of an equivalent of a maser working with visible light. Others, including Ali Javan, Donald Herriott, John Sanders, Herbert Cummins, Isaac Abella, Geoffrey Garrett, Paul Rabinowicz, Steve Jacobs, Irwin Wieder, Peter Sorokin, Wolfgang Keiser, Mirek Stevenson, Ron Martin, Valentin Fabrikant, Fatima Butayeva, Charles Asawa, Ben Senitzky, Elias Snitzer and William Bennett, would all make steps that proved helpful to the three main players. Though the individual names in this list are likely to be meaningless to the reader, it gives a feel for the degree of parallel working in an attempt to get there first, the feverish activity behind the scenes, some in the same establishments as Schawlow and Gould, while other teams laboured at universities and companies across the world.

Another name we have already met, working directly with Schawlow, was Charles Townes himself. After discovering the difficulties of simply moving existing maser technology to higher and higher microwave frequencies, he made the mental jump of thinking it might be easier to shift a good distance up the electromagnetic spectrum. In 1957 he began to consider what he would always refer to as an 'optical maser'. The title may seem a little clumsy, but the maser was his baby and he wanted to ensure that any future development would be tied clearly back to its roots and his device. Townes picked up on an idea called optical pumping, which was a way to use illumination with selected frequencies of light to push electrons in gas molecules up to a higher level to improve microwave transmission. Townes felt that this technique could also be used to start off the chain reaction of an optical maser.

Bell discoveries

Later that year, Townes moved from Columbia University back to AT&T's Bell Labs, where he would work with his brother-in-law Schawlow, forming one team in the laser race. Bell's greatness is something of a faded memory now, but the company's near-stranglehold on the US telecoms market meant that it could afford to throw a large amount of money at basic research and development. This was arguably the ideal place to work on an optical successor to the maser – after all, just the year before, it was Bell scientists who had won the Nobel Prize for their work on the transistor.

Townes, Schawlow and their team began, in a very careful, structured way, to examine the potential of various vaporised metals as substances to stimulate into emission. These initially seemed attractive as they were good candidates to get into the necessary excited state, but the practical side of dealing with metal vapours proved trickier than expected. At the same time, the team had to devise some kind of chamber for the stimulated emission to take place in. The microwaves in the maser had required only simple rectangular metal boxes with holes around the size of the microwave wavelength (a few centimetres) to let them escape. But it was hard to see how visible light, with its much smaller wavelength, could be suitably confined. You can't store light in a metal box.

It was Art Schawlow who came up with the solution that became pretty much universal in early lasers. He suggested using a Fabry-Perot cavity, a chamber that was open on the sides, allowing the light that would stimulate the medium to get in, and that had mirrors on each end, so that the emitted light barrelled back and forth between the reflective surfaces, constantly re-stimulating the medium inside. As long as the mirrors were highly reflective, exactly parallel and many wavelengths across, they should form the basis of an excellent chamber.

A gem of a laser

Up to now, all the ideas for optical masers had focused on gases (metallic or otherwise) as the material to produce stimulated emission, but Bell Labs' success with the transistor was a powerful reminder of the importance of considering the solid-state. Schawlow began to look into using some kind of transparent solid instead. Synthetic rubies were already used in masers, so this seemed an obvious starting point. With the systematic scientific viewpoint prevalent at Bell, this meant studying the different excitation states of the chromium atoms that are added to aluminium oxide to produce rubies, something that wasn't properly understood at the time. After some considerable effort, rubies were dismissed. It was difficult to get enough of the chromium atoms into an excited state, ready to be triggered, and the team's research suggested that rubies would absorb too much of the light energy, wasting it as heat rather than re-emitting it to get an effective stimulated emission reaction.

By the end of 1958, Townes and Schawlow had published a paper outlining their optical maser concept in *Physical Review Letters*, and, after a lot of grumbling from the Bell Patent Office, which found it hard to believe that an optical maser had any practical use, filed patents on the use of optical masers in communications. Almost inevitably, Bell Labs saw the world in terms of communications, because this was their speciality. The scientists working on optical masers considered them an exciting prospect because using the optical region meant you could pack far more signals into a 'pipe', provided you could produce light on a tight band of frequencies. This was exactly what the concept of the optical maser promised. As far as Bell was concerned, their scientists were well ahead of the race – but they had no idea of the competition that already existed.

The Gould standard

While at Columbia, Townes had asked for some help from a graduate student called Gordon Gould. They had discussed optical pumping and the use of a particular type of lamp. But Gould had already been thinking about an optical version of a maser, and the comments from Townes spurred him into action. They were very different characters: Townes the straitlaced, solid, establishment type – Gould more of a revolutionary, given to working in frantic periods of effort then doing very little for days at a time. Where Townes' background was unimpeachable, Gould had dabbled with Marxism, a youthful dalliance that would come back to haunt him.

Gould independently came up with the idea of using parallel mirrors to produce repeated passes of light through a medium, building excitation, and where Townes had merely seen the optical maser as a variant of his low-power microwave device, Gould realised there was a potential for pass after pass through the medium to build up more and more stimulation, eventually producing an intensely concentrated beam of light. He believed it should be possible to create a beam that would easily produce temperatures as high as the Sun's surface – around 5,500°C – or that could even compress atomic nuclei until they fused as they did in the massive temperatures and pressures of the Sun's interior.

The laser is named

It was November 1957. Gould knew the importance of establishing priority in such a competitive scientific environment. He wrote up careful notes in a form that could be used as evidence for a patent, conjuring up a new term for the device that would rapidly displace 'optical maser'. Gould headed his notes: 'Some rough calculations on the feasibility of a LASER: Light Amplification by the Stimulated Emission of Radiation.' The normal thing to do would be to have this document signed off by a colleague, but Gould was aware there was already competition at Columbia. Wary of losing his lead, he instead took his notebook to a notary public (by coincidence also called Gould), a low-level official in America who can administer oaths and witness documents. The notary stamped and dated Gould's pages to establish just when he had come up with the idea.

In principle Gould was now all ready to get his patent application in several months before Townes, but at this point he made a crucial mistake. His parents got him an introduction to a patent lawyer in January 1958, but Gould came away from the meeting with the incorrect idea that he needed to build a working model before he could be granted a patent. (Working models were required for perpetual motion machines for obvious reasons, but usually a patent could be granted on appropriate paperwork alone.) Gould had already worked briefly with a company called Technical Research Group (TRG) that specialised in defence contracts. He now got a job there working on atomic clocks, hoping to spend his lunchtimes and evenings on his laser project.

To begin with, Gould kept this to himself, but he was soon required to sign a document waiving his patent rights on inventions he made while working for TRG. He explained he needed an existing idea to be excluded from this waiver. The company agreed, and wanted to hear more. Initially they doubted the value of a laser, but Gould was a passionate advocate and managed to sell them on the prospect. TRG's founder, Lawrence Goldmuntz, suggested that building a laser could be a project that would gain Pentagon funding, meaning they could swing heavy-duty resources behind it, even though Gould would keep the patent rights. Gould piled on the possibilities, talking up a device that he said could send a tight communications beam all the way to Mars or punch holes in metal with the sheer concentrated energy of its beam of light. One day after the publication of Townes and Schawlow's paper, on 16 December 1958, the TRG proposal went off to the Pentagon, asking for a hefty $300,000 of funding.

Military might

Until this point in history it had been difficult to interest the conservative US military in leading-edge technology, but the proposal arrived at an opportune moment. In response to the shock of the Soviet Sputnik becoming the first man-made satellite, flaunting apparent Soviet technological superiority, the US government had set up the Advanced Research Projects Agency. ARPA was to fund and manage precisely the kind of crazy possibilities that Gould was claiming for his laser. Gould responded by dreaming up other potential military uses for his device, from acting as an ultra-precise radar, with its much shorter wavelength resulting in far greater precision than traditional equipment, to the science fiction vision of projecting a bright glowing spot on a distant person or piece of military hardware to help target incoming weaponry. Perhaps, Gould suggested, in a precursor to Ronald Reagan's 1980s 'Star Wars' Strategic Defense Initiative, the power of lasers could even be used to destroy incoming missiles. This possibility led to the Air Force giving the project the codename Defender.

Not only did Gould's enthusiastic and visionary ideas inspire the ARPA team, it resulted in a shocking outcome. Most companies are used to government agencies attempting to trim their budget. It's the kind of game everyone expects when bidding for government contracts. In response to the request for $300,000, ARPA came back with approval for a spend of practically $1 million on the laser project, requesting that TRG went into overdrive, testing different technologies in parallel rather than carefully working through the options one at a time. It seemed as if everything was about to take off for Gould – that it would be plain sailing from now on. But the military bureaucracy had no intention of letting that happen.

Your brain is classified

The powers-that-be decided that with all its potential military applications, the TRG laser project should be classified, a common enough move. Gould's first concern was that his potential for a patent – the whole reason that he came to TRG in the first place – would be scuppered, as you can't properly patent something that has restricted access. To get around this, TRG split the information on the laser into two, separating the underlying technology from the applications, enabling them to file a patent application in April 1959. Gould probably thought the worst was over. True, he had a history of dalliance with left-wing ideas that would inevitably be considered suspicious to an America that was still reeling from the McCarthy era, but he was assured by TRG's ARPA contacts that his security clearance should be simple. Unfortunately, the newly formed ARPA had not yet developed the expertise to master the complex military game of security levels.

Gould's security position came to a head when Gould and Goldmuntz had a meeting with RAND (an acronym for Research ANd Development), the think-tank set up at the behest of the US government in the 1940s. As Gould and Goldmuntz settled down to discuss Gould's proposal with RAND's scientists, an official took the document away, saying that Gould was not allowed to see it, as he did not have high enough security clearance. This, remember, was a proposal describing Gould's work and ideas. He wasn't allowed to look at his own proposal. Similarly, Gould was immediately restricted from entering the building at TRG where the laser work was being carried out. It is a powerful reminder of how different times were, that two of the genuine problems in obtaining clearance for Gould were that he had lived with his wife before they were married, and that two of the referees for his security clearance had beards. Sporting facial hair was considered a sign of being subversive. You couldn't make this up.

Rubies return

While Gould and the Bell Labs team were still struggling with the bureaucracy, and Bell's team continued with painstaking research, a third player entered the field. The Hughes Aircraft Corporation, founded by the reclusive Howard Hughes, was a major US defence contractor in the 1950s and like Bell was prepared to indulge in fundamental research in the hope of developing new products. One such project was work on an improved maser with the aim of using it to achieve more precise radar, and it was for this that the Hughes Corporation hired in 1956, among others, Theodore Maiman, a young man with an ideal combination of an electrical engineering degree and a doctorate in physics. Here was someone with both the theoretical and practical training to make the laser a reality.

Maiman learned a lot about working with rubies in a project exploring the potential of a ruby-based maser. This was a fiddly business – to work at the kind of energy levels required for microwaves, a ruby had to be cooled with liquid helium or liquid nitrogen – but it made Maiman very much the master of ruby manipulation. More enthusiastic about research than pure engineering, Maiman put a lot of effort into studying the way that the chromium atoms in the ruby could be pushed up to higher energy levels. He picked up on the optical pumping with visible light that had been written up by Townes and wondered if this would be helpful for a ruby maser … but it also got him to thinking about producing a visible light output.

Security follies

Meanwhile, the most likely contender to get there first, Gordon Gould, despite the help of corporate lawyers, was still fighting to get his security clearance. A key problem was Gould's reluctance to name former friends who had been fellow communists. His sense of loyalty made the paranoid administration suspicious that despite his capitalist veneer and willingness to work for the military-industrial complex, Gould still harboured overly liberal left-wing leanings. Unfortunately not all of Gould's old friends were as high-minded as he was. One, Herbert Sandberg, in an attempt to clear himself, named Gould as a known security risk. This bombshell came as it looked as if Gould would get the clearance, sending everything back to square one in the nightmare game of security snakes and ladders.

By now there were as many as twenty people working on the TRG laser project, supposedly Gould's baby, yet he wasn't even allowed into the building where they worked. They could not ask for advice from him directly, only by asking indirect, evasive, hypothetical questions. It was like trying to carry out a car service down a phone line. Things came to a head when Gould's notebooks were confiscated, because he didn't have the security clearance required to read the words that he had written.

Error correction

The opposition could have got there easily before Maiman, but Gould was hampered by the security issues, while Schawlow was convinced that rubies would not work because he believed they were too inefficient at converting the light energy that was used to stimulate them into coherent beams of laser light. This conviction was primarily based on an incorrect value for the efficiency of the process given by another researcher at a conference in June 1959, which Schawlow didn't bother to check. This small lapse was enough to lose him the race. As did Gould, Schawlow continued to focus on the better-understood but more operationally tricky use of gases and particularly metal vapours as materials for stimulation. Maiman had also heard about the problems with rubies from Schawlow at a conference in September that year, but he felt that there was something not right with Schawlow's concerns, given his own experience with the material.

Maiman had two particular issues. One was that Schawlow had said it would not be possible to get enough electrons excited in ruby because the crystal would be bleached by the strong light – but Maiman knew that in the optics of such crystals, bleaching was not about being washed out, but was a term meaning that most or all of the electrons *were* excited – exactly the right state for stimulated emission. This seemed a positive benefit rather than a reason for ignoring rubies. Maiman also had a problem with the efficiency calculations used to argue against ruby's effectiveness. Something seemed wrong, and it nagged at him. His superior didn't agree, persuaded by Schawlow's talk, but Maiman insisted that ruby was worth a go, and won the argument.

The thing that particularly frustrated Maiman was the efficiency problem. If most of the energy of the photons being pumped into the ruby did not end up producing stimulated emission, where was it being absorbed or being re-radiated? If you put energy in, it has to end up somewhere, yet no one seemed to know exactly where it was going. By now Schawlow had abandoned ruby at Bell Labs, believing that it would only possibly work if cooled with liquid helium, which proved an experimental nightmare. Maiman knew that ruby was a strong absorber of both yellow/green and blue/violet light, yet according to the figures from the conference, only about 1 per cent of this energy was resulting in stimulated emission.

There seemed to be two main ways that the energy could escape. One was by simple scattering, the effect that produces the blue sky and enables us to see objects. In this, rather than drop down to an intermediate level before emitting the red photon, a stimulated electron drops back all the way to its initial energy level and puts out the same energy of light that it absorbed. The other possibility for using up energy is that the energy from the electron is transferred to kinetic energy in the atoms in the substance. The material heats up. In trying to quantify how the energy was lost via these alternative routes, Maiman hoped to discover the kind of material that would work better than ruby in a laser. But instead he got a shock.

In his experiments, Maiman discovered that there was relatively little scattering or heating. His first guess at the actual efficiency of ruby was around 70 per cent, and it was later measured more accurately in the high 90s. Somehow, the opposition had got the key measurement wildly wrong and had based their entire strategy on this mistake. It didn't help at this point that Maiman's managers seemed more willing to believe the Bell Labs results than his own. Unable to get permission to proceed officially, he started work on a ruby laser without informing his boss.

A flash solution

The problem now was to get a bright enough light to illuminate the ruby and pump those electrons up, ready to be triggered. Arc lamps were bright enough, but generated too much heat, damaging the ruby, so Maiman's team turned to industrial versions of cinema projector bulbs. These were fiddly technology, requiring cooling systems to avoid them burning out, but they were certainly intense. All the ideas for laser design to this point had assumed that the laser would be a continuous beam, but Maiman's assistant, Charlie Asawa, also looked at the possibility of using pulses of light to stimulate the lasing material. This seemed appealing, as it gave the lamp a chance to cool between flashes to avoid overheating. And here was where one of those examples of serendipity that often come up in science occurred.

One of Asawa's friends was an enthusiastic photographer and had recently bought the latest snapper's gadget. Back then, night-time photography required flashbulbs, one-shot bulbs that burned out a strip of magnesium or zirconium in a high-oxygen atmosphere to produce an intense flash of light, then had to be thrown away. But Asawa's friend had bought one of the newly developed electronic flashes, which built up an electrical charge in a capacitor, then released it in one go across a low-pressure tube typically containing xenon gas, producing a blindingly bright flash that could be repeated as quickly as the capacitor could be recharged, without changing the bulb.

The light from an electronic flash seemed ideal in terms of the energy levels provided, was easily controlled and required none of the messy cooling mechanisms of the specialist projector-type bulbs. Admittedly, no one had thought of a laser producing its light in pulses rather than as a continuous beam – and it probably wouldn't be any use for applications like communications – but at least it would demonstrate the feasibility of visible light-stimulated emission. Maiman decided to give it a go – only to have the Hughes Corporation put an unintentional spanner in the works.

In early 1960, the company moved their labs from Culver City to a new location in Malibu. While this might have been an attractive lifestyle relocation, it proved highly inconvenient timing, as just when the prize seemed in sight, Maiman had to pack up his lab and move. Luckily for him, both Schawlow and Gould were hitting practical problems with their gas lasers, suffering everything from metal deposits turning the tube containing the glass opaque to difficulties getting an appropriate light source to do the pumping. Gould was also still hands-off, unable to contribute directly to his own project. He had sat through a security hearing at the Pentagon in April, but it proved inconclusive.

Lasers go live

The same month as Gould's hearing, Maiman was finally able to get started on a practical ruby laser. He made use of the then common spiral form of flash tubes, slipping a thin cylinder of ruby inside the spiral. Each end of the ruby rod had silver mirroring, but one end had a small hole scraped in the silver so that the laser light could emerge. The whole thing was enclosed in a reflective aluminium casing, which both protected the scientists' eyes from the intense flashlight and reflected light back inwards, ensuring that the maximum amount of the flash was directed in towards the ruby rod. Compared with the large, clumsy, bench-sized apparatus required for the gas laser experiments, the whole thing was smaller than a fist: compact and neat.

With an experimental design constructed, there was only the need to see if it worked. This wasn't as easy as it sounds. Rubies fluoresce naturally with a red light that is not the coherent, tight beam of stimulated emission. Just because the device produced a red light did not mean it was lasing. So Maiman had to test for a sudden sharp spike in a tight frequency range of a spontaneous emission cascade, rather than the lazier, spaced out glow of ordinary fluorescence. By 16 May 1960 he and his team were ready to give it a go.

The procedure was far more like a Hollywood version of a scientific experiment than is usually the case. Maiman started off with a relatively low voltage on the flashlight and produced a red spot of conventional fluorescence on a white target. As he and his team gradually cranked the voltage up, making the light more and more intense, the spot got gradually brighter. But suddenly, with one more increase in voltage, the output spiked, the length of the flash measured on an oscilloscope collapsed and the intense red spot reflected from the screen lit up the darkened room.

Success had come from an unexpected direction. It wasn't all over on 16 May, though. Maiman had to get his results published, and was shocked to get a rejection from *Physical Review Letters*, in part because he had made the mistake of referring to the device as an optical maser, and the *Review*'s editor had decided by then that masers were old news. Instead, Maiman managed to get a short letter into the prestigious journal *Nature* and scheduled a longer paper (that would eventually be published elsewhere) with the slow-moving *Journal of Applied Physics*. Aware that time was ticking by, the Hughes Corporation also arranged a press conference, which on 7 July brought the laser to the attention of the world's press.

In terms of presentation, this event was solidly focused on the down-to-earth potential applications of a laser like enhanced communications, but the press knew that what they were seeing had the potential for a dramatic headline and pushed for a comment on whether this was a prototype ray gun. Maiman tried to put this off, but was unable to say that it would never be used as a weapon. This was enough for the press, and to Maiman's horror the headlines blared: 'L.A. man discovers science fiction death ray.'

Who got there first?

There was no doubt that Maiman had made the first laser, though there was a certain amount of carping from Bell Labs (stubbornly still referring to the device as an optical maser), who at one stage claimed that the ruby laser was their idea, despite Schawlow having dismissed it as unworkable. A Bell team did, however, manage to get the first continuous laser running, based on a helium/neon mix (eventually to become the technology used in the first supermarket scanners) just before Christmas of the same year.

As for the patent battle on the underlying concept of the laser, Gould (who never did receive his security clearance) managed to get a UK patent in 1964, but he had to wait until the 1980s before his patent right was acknowledged in the United States. Bizarrely, the Nobel Prize awarded in 1964, which confusingly and clumsily referred to the 'maser-laser principle', went to Townes, Basov and Prochorov. Despite the predilection of the Nobel committee to award prizes to theorists rather than experimenters, this still seems an odd selection of names to choose, as it is clearly oriented to the maser rather than the much more significant laser.

A new kind of light

All the light that humans had experienced to date, whether from the Sun or a heated object like a flame or the filament in a light bulb, was chaotic. Photons flew off in any old direction with their phases unconnected, and came in a whole mix of energies and hence of colours. But the waterfall triggering of the laser and maser meant that the photons were in step, a beam of near-monochrome light that was highly directional. The beam from a laser is so tightly in step, so difficult to scatter, that laser beams have been bounced off the Moon and returned to Earth still in a relatively tight bundle of photons.

As became obvious after Maiman's press conference, as soon as lasers were publicised, thoughts turned to how this new type of light could be used. If the development of the laser was a remarkable achievement in itself, the way lasers would be exploited took this light – the hallmark of the quantum age – into whole new realms.

CHAPTER 7

Making light work

It was inevitable that at the press conference to announce the first laser, the press corps would think of that science fiction staple, the ray gun. Here was a very powerful beam of light that surely would soon be made into a weapon. Only a remarkably short period of time, just four years, elapsed between the first experimental laser being produced and James Bond's archetypal experience in *Goldfinger* where the ruby red beam cuts through a block of gold as if it were butter, heading inexorably towards splitting the immobilised Sean Connery in two.

As it happens, by far the majority of uses of lasers would not involve such dramatic activities, but there is no doubt that lasers can pack a punch. So how does a beam of insubstantial light manage to cut through gold and slice up spies? It's worth taking a step back to something most of us have played with as a child – using a lens as a burning glass.

Heat from light

We are used to the feel of sunlight's warmth on our skins. One of the forgotten senses, when we claim there are five, is the way that our skins can detect infrared, light that is just below the visible range in the spectrum. But focus the rays of the Sun with a convex lens and the result is far from a pleasant warmth – in fact there is enough concentrated energy to scorch wood and make paper burst into flames. Some of the incident photons are re-emitted, enabling us to see the object in the sunlight, but when others excite electrons, the resultant energy goes into vibration in the atoms: it heats them up. By concentrating the light with the lens there is simply too much energy pumped into too small a region. The temperature shoots up to the extent that the material starts to react with the oxygen in the air in the chemical process we call combustion.

The burning glass rarely manages much more than setting a piece of paper alight, although house fires are sometimes caused by glass vases focusing the Sun's rays and starting a blaze. But way back in Ancient Greek times, the mathematician and engineer Archimedes proposed an optical defence mechanism for his city. In danger of attack by Roman vessels, he suggested that huge curved mirrors be constructed on the harbour. These would be used to concentrate the rays of the Sun, setting the attacking ships alight. The principle might well have worked had the ships been still enough, especially if tar had been used to seal gaps in the wood, though as far as we aware the mirrors were never constructed. In principle, though, here was the first death ray.

A laser cuts on exactly the same principle as a burning glass, but with the difference that the light is far more concentrated, even at a considerable distance, and the waves (or phase of the photons) are in step, making the impact more likely to transfer energy to the target efficiently. Whether it's melting gold or operating on an eye, the laser raises the temperature in a small, controlled region extremely quickly, providing the desired cutting effect. The question that interested the military, of course, was whether this ability to cause damage could be imposed at a distance, turning the laser into a formidable weapon. The answer would be yes ... and no.

Death rays

As Gordon Gould suggested, an almost inevitable military application of lasers is in targeting weapons – either as a visual cue, where a conventional gun or rifle projects a laser dot on the target, or used to guide bombs to a location pinpointed by an infrared laser. But when it comes to using lasers directly as weapons, we are still a fair way from the classic sci-fi ray gun. Probably the closest to the handheld ray gun have been attempts to use lasers as a device to temporarily blind an opponent. There is no doubt that direct exposure to a laser can cause temporary blindness, but the borderline in eye damage between temporary and permanent is a fine one and these weapons are widely considered unethical, resulting in the development of the United Nations 'Protocol on Blinding Laser Weapons' which has been active since 1998.

The protocol, however, only prevents the use of weapons specifically designed to cause *permanent* blindness, a loophole that means in practice there are still likely to be weapons deployed with the aim of causing a temporary effect, whether or not that may sometimes result in permanent blindness. (There is something of a parallel with the use of tasers, which are deployed as non-lethal weapons despite Amnesty International's claim that over 500 people died as a result of their use between 2001 and 2012.)

To have the kind of destructive power we expect from a good science fiction ray gun, current lasers need to be far too big to be carried as a sidearm. Mostly as yet these are still under development, but there are examples that are close to being deployed like the US Navy's 'Laser Weapon System', a ship-mounted device intended to use an infrared laser to cripple drones and small boats. Other US concepts include airborne lasers designed to make surgical strikes on ground targets or destroy missiles. Such weapons would not work by outright disintegration but either by heating the outer skin of a target, causing stress failure, or vaporising some of the surface sufficiently aggressively to cause damaging shock waves. At the time of writing, most attempts at producing laser weapons have been cut as a result of budget reductions. There is a fair way to go before a laser could be an easily deployed battlefield weapon.

Fusing with light

The idea of causing stress by vaporising the surface of a target may seem extreme, but it is an approach that is being used in one of the most dramatic applications of lasers: nuclear fusion by inertial confinement. Take a visit to the National Ignition Facility at the Lawrence Livermore Laboratory in California and you will be faced by a laser set-up that would make any Bond villain proud. In two vast ten-storey halls, a tiny initial beam undergoes a transformation that will make it into a monster.

A small triggering laser's infrared output is split 48 ways before each sub-beam is passed through an amplifying laser that boosts the beams' power by a factor of 10 billion. Each of those beams is then split again, producing a final 192 beams that then pass through the vast main amplifiers, adding another factor of a million to bring the overall power up to a sizzling 6 megajoules. (The flash is so powerful that for a few trillionths of a second it is as if the output of 5,000,000,000,000 traditional light bulbs was concentrated into a tiny but immensely powerful flare of coherent light.)

These 192 beams are up-converted to ultraviolet, which is better suited to its final task. In a reaction chamber the beams converge on a small frozen pellet of deuterium/tritium fuel. At the moment of impact, the outer surface of the pellet is vaporised, producing an intense shock wave that compresses the remaining fuel to such an extent that nuclear fusion can take place. At the time of writing, this was yet to achieve 'ignition' – where more energy is produced than is required to produce the fusion – though the National Ignition Facility has managed to yield more energy than is applied to the target. (Lasers aren't 100 per cent efficient, and most of the energy consumed by the device is wasted in the amplifiers rather than applied to the pellet.)

Lasers, lasers everywhere

While both gas and crystal lasers are common enough in industrial and military applications, and turn up in the familiar supermarket checkout scanners, neither of them features in the lasers we all are likely to have around the home – the ones in laser printers, laser pointers and players for CDs, DVDs and Blu-ray. These are all semiconductor lasers, in the same family as the LED lighting that is taking over as a low-energy standard. The first thought that it might be possible to produce laser light from a semiconductor goes all the way back to 1958, before any lasers had been constructed, and it was as early as 1962 that a visible laser based on a semiconductor was constructed. This was not a true predecessor of modern solid-state lasers as it could work only in pulses and at cryogenic temperatures, but by 1970, in parallel developments at Bell Labs and in the USSR, the modern room-temperature semiconductor laser was developed.

This new form of laser made use of a structure known as a 'heterojunction' – an interface between thin layers of semiconductor, which have unequal gaps between their valence and conduction bands. In the case of the semiconductor laser, this took the form of a narrow band gap material sandwiched between two wide gaps. There would be a whole family of different variants, each effectively a special version of a light-emitting diode, where the structure enables the special 'cascade'-like properties of the laser, producing coherent light, typically by surrounding the diode with some kind of optical cavity. Not only are such devices much cheaper than the traditional types of laser, they can be made much smaller to fit in a wide range of equipment. Semiconductor lasers outsell traditional lasers by a factor of nearly a thousand, and are likely to be present in most homes and workplaces.

True view

Apart from threatening secret agents with being sliced and diced, another early application of lasers that has taken longer than expected to become useful is the hologram, which was invented in principle before the laser even existed. The hologram was dreamed up by the Hungarian-British scientist Dennis Gabor, soon after the Second World War. Gabor was never one to consider pure theory, always wanting a practical application for his ideas. He had built a do-it-yourself X-ray machine in his teens, but though he had originally studied engineering in Berlin, the influence of world-class physicists had driven him across to focus on science.

Gabor was thinking of ways to improve on the electron microscope, which at the time was in its infancy. He realised that a microscope's image could be enhanced if it could somehow provide a wider viewing angle. This particular application proved impractical, but it started Gabor on the path to devising a way to take a photograph, using light, that could be viewed just like a real object, seeing different perspectives from different directions, rather than just a flat picture.

The secret behind the hologram, Gabor's idea of a truly three-dimensional image that changed as you looked at different angles, was thinking about how looking at a flat photograph and looking at a real scene differs. If you imagine having two windows side by side, one with a perfect photograph of a cityscape in it, so high in resolution that it is indistinguishable from the real thing, and the second a normal glass window, you can easily see the difference. As you move around, parallax caused by objects being at different distances behind the window means that they seem to move with respect to each other. When you go over to one side of the window, for instance, you can see an object that was previously behind another, because the light from the object is no longer screened by the item in between. This doesn't happen with the flat photograph.

Now think of the surface of the glass window. Every single photon from those different objects, travelling at different angles, is arriving at the surface of the glass. All the photograph does is to sum up the intensity and colour at a particular point. But imagine that instead you could capture information on every single photon, then take the window elsewhere. If you could stimulate the window to start producing the same photons again, then you should have a true three-dimensional view captured on the flat surface of the glass.

This is how a hologram works. To simplify things, imagine that instead of a cityscape we have a small child's toy, and it is illuminated with a monochrome light (we'll make it green, to match the early holograms). The light reflected from the toy hits our window glass.

Each photon has not just an intensity and a direction, it has a phase – but only the intensity is captured if we use photographic film. But imagine instead we shone a second beam of light onto the surface of the glass from the same direction. The light from the toy and the direct light would interfere, producing an interference pattern. If we could store that pattern, it would hold far more information than a normal photograph. And if we could use that pattern to recreate the stream of photons emerging from the glass, we would recreate the three-dimensional image.

This was all hypothetical as far as Gabor was concerned, because to make it work, those photons would have to be very special, linking in phase and of identical energy, otherwise the interference effect would not produce the desired result. Good filters would make a relatively monochrome light, but there was nothing that could be done about the phases of the photons – until a laser came along and delivered coherent light with all the photons triggered in step. It was only four years after the first working laser became available that Emmett Leith and Juris Upatnieks at the University of Michigan pro-duced the first true hologram, a bizarre still life of a model train and a pair of stuffed pigeons. Gabor's impossible vision had been made possible.

I remember visiting one of the first big exhibitions of holograms, called 'The Light Fantastic' and shown at the Royal Academy in London in 1977. In one sense it was very mundane. It was like looking through a series of fuzzy little windows onto objects illuminated with a sparklingly bright green light. But at the same time, in a sort of mental quantum superposition, you realised there was no object there. And then the laser would snap off, and you would see just how much this apparently three-dimensional thing was an illusion. All you were looking at was a piece of glass with a meaningless pattern of speckles on it.

A shattered illusion

Thinking of my original 'picture alongside window' experiment gives us another insight into the remarkable nature of holograms and how they differ from a trad-itional photograph. Imagine I blanked out all but the top left segment of both 'windows', leaving just a square two centimetres on each side visible. As far as the photograph is concerned, I have lost almost all the information it held. If, for instance, that top left square showed only blue sky, then all I would see would be sky. I could tell nothing about the cityscape from that tiny segment.

However, things are quite different with the real glass window. Yes, standing in the same relative position to this as the cameraman was to the photograph, I will see only a square of sky. But if I get up close and move around, I can look through the square of window at different angles and still see most of the cityscape. I won't have as a good a three-dimensional view because I can only see through this small segment of glass, but by looking at an extreme angle I can take in pretty well all of the scene. The same must be true of the hologram. If I break off just the corner of the hologram, unlike the traditional photograph, I still have information about the whole scene.

There will be fewer photons hitting the small square of hologram, so the image will be fainter and less clear (there are also limitations due to the resolution limit of the medium used to capture the image), but the fact remains that unlike the traditional photograph, any piece of a hologram will contain vastly more information, enabling the viewer to reconstruct the whole image that was visible.

On reflection

It's fair to say that holograms were, even more than lasers, first seen as a technology in search of an application. In practice, the most valuable applications of holograms have gone in two very different directions – the simplistic holographic strips used for security protection, and holographic data storage.

The most common security holograms, the sort of things you see on credit and debit cards, some fancy bank notes and on premium products, are usually reflection holograms, where the interference pattern of a holographic image has been stamped on a metal foil. Even though this doesn't produce a true holographic 3D image, it makes use of a clever trick so that the viewer does not just see a speckled interference pattern.

A reflection hologram has two or more layers. At each layer, some of the light is reflected back, and it is the interference between these different layer reflections that makes the holographic image emerge from what otherwise would just be a multi-coloured mess. It is sometimes said that these security tags aren't holograms at all. That's a bit unfair – they are, but they don't provide the true 3D image that allows you to see an object from different directions, as if you were looking at the real thing. It is holographic technology, but not producing a traditional visual hologram.

Some, however, are a special variant of the true hologram that can be viewed using white light, called a rainbow hologram, for the obvious reason that the image appears to have an unnatural range of rainbow colours. The holographic image is produced in the usual way, but through a narrow slit. When light is sent through the hologram, different parts of the image are seen depending on the wavelength of the light – so white light produces a whole image, but with different strips of it in different colours. Like most holograms, rainbow holograms need to be lit from the back – to make them work as a security tag, they are backed with reflective material so light from the front passes through the hologram, then back out through it after hitting the reflector.

Three-dimensional memory

There is no doubt that holographic security tags are widely used and valuable, but they aren't the sort of breathtaking application we would hope for from such a dramatic piece of quantum technology. However, another way of using holograms holds out more promise: holographic data storage. As our computers become more and more powerful and ever-present (remember, the phone in your pocket is a sophisticated computer in its own right), we also need ever-greater quantities of storage. With cameras always ready, many of us will take hundreds or thousands of photographs a year, where once, using film, we might have taken a few dozen.

At the same time, our storage requirements for music and video are going through the roof. For a while, the laser provided one of our most common storage mechanisms, burning markings into writable CDs and DVDs in optical storage, but these are now on the wane. Just as the once ubiquitous diskette drives have disappeared from our PCs, so manufacturers are increasingly dropping optical drives as we make use of the internet to store most of our material in 'the Cloud'. But the Cloud is only a conceptual entity. Behind this pleasantly fuzzy notion of our documents and pictures, videos and songs disappearing into a white fluffy mass is a reality of massive data centres, where the information is stored on old-fashioned magnetic disc drives. Although our portable devices often hold information in solid-state memory, the data centre's magnetic discs – much cheaper to make on a large scale – are the only viable solution for the vast quantities of storage required.

However, that storage requirement is always on the increase, taking up ever more space and power. When, for instance, a photographic Cloud site like Flickr gives each of its users 1 TB – a million megabytes – of free storage, it is inevitable that those data centres will end up creaking at the seams. Holograms have a potential answer, especially for the sort of information that is rarely changed, like copies of songs and videos. This is because a hologram is really just a very compact way to store a lot of data. Usually this is the complex information required to reconstruct the three-dimensional image, but there is no reason why it can't be used to pack away traditional digital data.

The concept behind holographic storage, which is still more theory than practice, is that rather than have just a single plane of interference patterns, as we experience in a visual hologram, the store will be a three-dimensional crystal, storing information on thousands of planes within the material, effectively stacking holograms many deep. If such an approach can be mastered, there are three huge advantages over a traditional disc drive. The most obvious is that the crystal is much more robust. A hard disc involves floating a head less than a hair's width, moving at the speed of a Boeing 747, over a very delicate platter. It's no surprise that they are fragile. But a holographic crystal has no moving parts within the storage mechanism.

The second benefit is sheer capacity. A magnetic disc is two-dimensional. Although the drives are very slim these days, they can't rival the way that a holographic crystal could stack thousands of layers into the same space as a single disc drive. And then there is speed. A disc is limited by the time it takes the head to move to the appropriate position, and then by the need to respond to magnetic patterns on the disc's platter in a linear fashion, bit by bit. In a hologram, the data can be accessed in parallel, pulling in far more at a time, and can be switched around at the speed at which a beam of light can be displaced.

The downside of holographic storage is that it is much slower to write to than is a magnetic disc. So once the significant technical issues in making holographic storage practical, robust and commercial are overcome, it is likely that a data centre will be mixed-mode. Magnetic discs will still be used for the volatile, quick-changing data. But in the background, data can be shifted off to the holographic store, where they can then be accessed with lightning speed. A document you create, edit a couple of times over a few hours and then delete will probably never make it to the crystal. But if I look at my disc usage, the vast majority of documents have not been edited in weeks, months or years; plus there's the much larger space taken up by the likes of photos, music and video. All these could be conveniently shifted off expensive, bulky magnetic drives onto holographic crystals, were they available.

A real world view

What we perhaps were expecting in the early days as just a matter of technological development – something that would come with time – were true holographic photographs. Images in a newspaper, say, or on a computer screen or in the family album that had all the 3D crispness of reality, with none of the monochrome fuzz of the early holograms nor the cardboard cut-out oddity of traditional

3D photography, a process that dates back to Victorian times and relies on putting a separate image in front of each eye. There is a third dimension in such photographs (or in 3D movies) but they could never be mistaken for a real three-dimensional world.

The chances are, at least unless and until there is some massive breakthrough in holographic technology, that we are not going to see any such development. Lasers are inherently monochromatic and the hologram's need to have an image illuminated by laser means that no current-generation hologram could be captured in natural light or carry natural coloration. There is no obvious leap to be made that would take holograms into the world of realistic portrayal of the world we see – and though science fiction often represents moving 3D images as being some form of advanced hologram (think of the Princess Leia image projected by R2-D2 in the first *Star Wars* film), in reality the chances are that any such future ability to project a three-dimensional image into empty space will depend on a totally different technology.

A glittering Diamond

Before leaving the strange and wonderful world of quantum light, there is one light source that is very different from a laser and that is doing remarkable work in the Oxfordshire countryside. Pretty well everyone has heard of the Large Hadron Collider (LHC) at CERN in Geneva. The Harwell-based Diamond Light Source resembles a miniature version of the LHC, but in terms of valuable contributions to the quantum age, it makes CERN's work look like yesterday's news.

Accelerators like the LHC push particles to near the speed of light. The work they undertake is usually described as particle physics rather than quantum physics. The distinction is a bit like the difference between zoology and biology. Quantum physics, like biology, gives us the fundamentals of how quantum particles behave, while particle physics gives us the details of the particle zoo, just as zoology tells us about specific animals. But all those particles studied by particle physicists are quantum particles.

Accelerators have been built since the middle of the last century to undertake a particularly brutal kind of science. They blast particles into each other at extremely high speeds and see what comes out. It's a bit like working out how a mechanical clock works by hitting it with a sledgehammer and filming how all the parts fly out of it in slow motion. The result is to give us access to particles that would otherwise never be seen. And early on it was discovered that these vast machine laboratories had an unwanted side effect. When the early accelerators known as synchrotrons were set up in the 1940s they were discovered to pump out large quantities of electromagnetic radiation. They emitted light.

Synchrotrons use a series of powerful electrical fields to accelerate particles, often electrons, which are steered around a loop by electromagnets. To get around the loop, the electrons had to be accelerated. This is because acceleration is a change in velocity, and velocity comprises both speed and direction. Even if something continues at the same speed, if it is pushed into a circle it is being accelerated. And, as Niels Bohr discovered when trying to uncover the structure of the atom, when an electron is accelerated it pumps out photons. If it weren't for the constant input of the synchrotron fields, the electrons would rapidly lose energy.

For the early researchers this 'synchrotron radiation' was a waste product, an unwanted side-effect. But just occasionally waste products can be useful. Think, for instance, of Marmite. When beer is made it leaves behind a gunky, sticky, dark residue. Rather than throw it away, the brewers found that this substance could make an interesting savoury spread – and Marmite (along with Vegemite in Australia) was born. What should have been a waste product became something valuable (if you like Marmite) in its own right.

The same thing happened with synchrotron radiation. The light produced by the accelerating electrons was available in a wide spectrum and was extremely intense, and eventually synchrotrons were built with the sole purpose of generating these bursts of radiation. The first in Britain was at Daresbury in Cheshire, replaced by the more sophisticated Diamond at Harwell. Apart from size and flexibility, the biggest advance Diamond has over its predecessor is containing undulators and wigglers. These are series of alternating magnets, which force the electrons into a pattern of sinusoidal ripples. Undulators produce a tight, narrow oscillation generating a narrow band of radiation, while wigglers produce a wider band of frequencies.

Around the main storage ring at Diamond are ranged beamlines, exit beams for the radiation where work-stations known unromantically as hutches house the experiments. In Diamond's massive 45,000-square-metre floorspace (around eight times that of St Paul's Cathedral) there are currently over twenty beamlines, with space for 40 in the final configuration. Depending on the position on the ring, these can produce infrared, visible light, ultraviolet or X-rays.

Inside the black box

Some of the applications of a synchrotron light source like Diamond are simply 'like the rest but better'. X-rays, for instance, have been used to determine the structure of crystals ever since the British father and son team of William and Lawrence Bragg won the Nobel Prize in 1915 for devising the technique. It's a direct application of a quantum phenomenon to uncover a structure via a distinctly indirect route.

Imagine you had a featureless box about the size of a shoe box, peppered on the outside with lots of holes. Inside the box is a structure you want to discover, but you can't open it. What you could do is drop a ball bearing through a hole and see where it comes out. Repeat the process with the box held at all sorts of different angles, and using all the different holes, and you could start to build up some kind of picture of what's inside. X-ray crystallography is a bit like this.

A beam of X-rays is shone into the crystal from many different angles. Inside, the X-ray photons will be absorbed by electrons on the atoms and re-emitted. These re-emitted photons will then interact with others from different positions in the crystal lattice. Depending on the spacing of that lattice, the phases of the photons will either reinforce or cancel each other out, so when the X-rays eventually emerge they will form a pattern of dark and light markings. By combining thousands of these images taken from different directions it is possible to deduce the structure of the crystal.

This was exactly the same technique used by Rosalind Franklin to produce the DNA diffraction patterns that would be used by Crick and Watson to deduce the structure of DNA. But as worked in the lab it is a slow process. A good example of how Diamond can make a difference comes from work by Pierre Rizkallah (a crystallographer) and David Cole (a biologist) from the University of Cardiff. The Cardiff pair were working at Diamond in 2013 on T-cells. These are white blood cells, the medical police of the bloodstream, responsible for destroying bacteria and other invaders.

T-cells have a unique ability to look inside another cell and see what's going on inside. They do this by using specially shaped proteins on the surface of the T-cell called T-cell receptors. These latch on to other large molecules call MHCs (major histocompatibility complexes). MHCs protrude from the outside of cells in the body, and differ depending on the internal make-up of the cell. A T-cell can distinguish between a friendly cell belonging to the body and an invader by the shape of the MHC and how well it fits in one of the T-cell receptors.

Sometimes, though, this process doesn't work correctly. The T-cell can attack friendly cells it doesn't recognise, as when someone has a transplanted organ. Or it can fail to spot a cell it could destroy but doesn't, typically a cancer cell. The MHCs on a cancer cell are too close to those of a normal cell for a T-cell to spot the difference. What Rizkallah and Cole are doing is using X-rays to study the shape of the T-cell receptors and how they interlock with the MHCs, so they can modify the receptors to latch on to a cancer cell but not the equivalent healthy cell. If this can be achieved, all they will need to do is extract some of a patient's T-cells, modify them and re-inject them at the cancer site to have targeted cancer killers.

The problem they would traditionally face is that in the lab it can take around eight hours to take a single exposure, because the complex structures make for a very faint image. As it takes thousands of exposures to build a single structure it could take up to three years to analyse one molecule. At Diamond, the X-rays are 100 billion times more powerful than an X-ray tube. The result is to provide such high resolution that images were initially produced in minutes, and now can be taken in a fraction of a second. The team can now analyse three structures a day and are hoping for even greater speed in the future.

Zapping poo

Some of the other applications of Diamond are more uniquely due to its optical power, with output millions of times brighter than the Sun. For example, in 2013, Mark Hodson from York University was using Diamond to study earthworm poo. Specifically, he was studying tiny calcium carbonate spheres around 2 millimetres across that are deposited in worm casts. Calcium carbonate is a very common mineral – it forms limestone, marble and chalk, for instance – and is used in everything from cement to paper whitening. Usually calcium carbonate has a crystalline structure, but in these worm droppings it is also in a non-crystalline amorphous form.

This is of real interest to material scientists because the amorphous form is usually unstable, collapsing into a crystal within minutes. But in the worm poo it can last for years. Some impurity is giving it extra stability. If this process could be understood it could be useful in everything from reducing the build-up of limescale in pipes to modifying the strength of building materials. In the lab it's possible to grind up the granules and identify the composition, but it is impossible to determine how the impurities are distributed through the amorphous material to keep it stable. By blasting it with powerful infrared light from Diamond they can obtain enough accuracy (the brighter the light, the better the resolution) to determine the worms' secret.

These are just two of many applications. Diamond runs 24 hours a day, blasting out light into the hutches (some bright yellow to warn that they are the lead-lined home of intensely powerful X-rays), allowing many different experiments to go on simultaneously, each making use of quantum light effects to gain better insights into the structures and nature of everything from integrated circuit designs and aircraft engine parts to the kind of natural structures we've already seen.

We are never going to see synchrotron light sources around the house, but lasers are here to stay. One of the essentials of moving the laser from being a specialist laboratory tool like Diamond to something that would have ready use in the world around us was the ability to use it at room temperature, rather than in a super-cooled cryogenic environment. But in the next chapter we will take a chilly plunge into a quantum phenomenon that was only discovered when working at the extremes of low temperature.

It was in the frigid region close to absolute zero that superconductors were first discovered.

CHAPTER 8

Resistance is futile

If you remember any physics from school over and above Newton's laws, you might recall a formula dealing with electricity: V=IR. The voltage (V) in an electrical circuit is equal to the amount of current flowing (I) times the resistance (R). Resistance is the electrical version of friction. Friction resists the movement of the object as it rests on a surface. Air resistance has an equivalent effect on a plane or a ball in flight. Similarly, when free electrons are flowing through a piece of metal they should just keep going for ever, but electrical resistance prevents this from happening. However, just over 100 years ago, the Dutch physicist Heike Kamerlingh Onnes noticed something quite remarkable.

Doctor Cold

Kamerlingh Onnes was a professor at the University of Leiden, primarily interested in cryogenics, the study of materials at very low temperatures. He was probably the world's leading expert on the mechanisms for producing seriously cold substances. (As I am going to mention him a lot in this chapter, I am going to take the liberty of shortening his surname to Onnes, although we really should use the whole 'Kamerlingh Onnes'.) In 1908 he managed to get liquid helium down to 1.5 K, the lowest temperature ever reached at that date. This is a temperature on the Kelvin scale, which is more convenient down at extreme low temperatures.

The Kelvin scale reflects the existence of absolute zero, the lowest temperature. This limit exists because the temperature of a substance is a measure of the energy of its atoms or molecules, and at absolute zero every atom would be in its lowest possible energy state. (In practice absolute zero can't be reached, because atoms are quantum particles that can never be pinned down with absolute certainty, as we've seen.) When absolute zero was first predicted at the end of the 17th century no one knew about atoms and molecules, let alone quantum theory, but they observed that the pressure of a gas fell as its temperature reduced. Absolute zero was the temperature at which it was predicted that pressure would cease to exist. On the familiar Celsius scale, absolute zero is −273.15°C, but when dealing with temperatures near to absolute zero, the Kelvin scale conveniently has the same size units as Celsius but starts from 0 at absolute zero.

The units of the scale are kelvins, represented by K (not °K), so 1.5 K is the equivalent of −271.65°C. In 1911, Onnes was experimenting on the conductivity of metals at these very low temperatures. When he got a piece of mercury down to 4.2 K (the familiar 'liquid metal' element is solid at these extreme temperatures, as it freezes at a balmy 234 K), its resistance disappeared entirely. It was like taking a moving object into a vacuum. With nothing to stop it, no friction, no air resistance, the object would keep moving for ever. Similarly, the electrons in the mercury acted as if there was nothing to stop them. There was no apparent resistance, they just flowed on indefinitely. If you set an electrical current going in a ring of such 'superconducting' material it would simply flow unchecked for as long as the low temperature was maintained.

Having a suspicion that there was zero resistance was one thing – demonstrating that it was exactly and not just approximately true was another. In practice it isn't possible to definitively prove that the resistance is totally zero; all that can be done is to establish that it is as nearly true as it is possible to measure. Even that is non-trivial. One way that Onnes attempted to show this was by setting a current going in a loop of superconducting wire within a bath of liquid helium. He then took magnetic field measurements outside the vessel over time and watched for a change – at its simplest, his experiment could be imagined as using a series of magnetic compasses, surrounding the vessel, watching for twitches in the needles. If the current in the loop was decreasing over time due to electrical resistance, the magnetic field it generated should change – but nothing happened. Onnes was able to perform this experiment only for a few hours before his helium evaporated away, though by the 1950s a similar experiment was run for eighteen months without any detected change in magnetic field due to a reduction in the current.

Challenging authority

Back in 1911, getting anything down to such low temperatures was a challenging exercise. (It still isn't the sort of thing you can do in the kitchen at home.) Take a look at photographs of Onnes' laboratory and you will see a steampunk version of CERN with great brass devices, large, important-looking gauges and a spaghetti-tangle of metal piping. In this laboratory equivalent of a ship's engine room, Onnes seems to have wielded the same power as a ship's captain. He had plenty of assistants, but you wouldn't realise this from his papers, where he was usually the sole author. (Admittedly this was a common failing among lead scientists then.) Even by the standards of the time, he was considered paternalistic and overbearing, an old-fashioned remnant of the traditional methods and modes of science that were being swept away by the new scientific class, an approach driven by sheer intellectual capability rather than social standing and authority. That's not to say that Onnes was anything but a superb scientist, but his attitudes were planted firmly in the previous century.

Not surprisingly, the discovery of superconductivity was more than a little startling both for Onnes and the physics community. It was fine for theoretical models to deal with a lack of resistance, but the real world was gritty. Resistance was a fact of life. Apart from anything else, superconductivity was the total opposite of what was expected to happen. By 1911, physicists were quite comfortable with the concept of the electron, a particle that carried electricity, the charge of which had just been pinned down by the American physicist Robert Millikan. It was thought that electricity was carried by a flow of electrons through a wire, acting like a gas being pumped through a pipe. It seemed reasonable, therefore, that if the temperature was low enough, the 'gas' would freeze, stopping the electrons flowing. Many expected the electrical resistance to shoot up towards infinity, rather than to disappear. Onnes himself had a different view that the resistance would drop as temperature fell, but that it would never make it to zero, as absolute zero was impossible to achieve. For him, it wasn't so much the fall in resistance that was surprising, but the sudden drop to zero at 1.5 K.

Electrical resistance, while highly useful in some electrical circuits, is often the bane of an electrical engineer's life. Think, for instance, of the way that we transmit electricity over large distances. The reason we need those vast unsightly pylons, carrying extremely high voltages, is that the higher the voltage, the less the transmission loss due to resistance. A fair percentage of the electrical energy in a conductor is converted to heat as the electrons interact with the atoms in the metal (think of an electric fire element as an example of this). With a superconductor there would be no transmission loss. A superconducting electricity grid would revolutionise the distribution of power. But it didn't take Onnes long to realise that this was little more than a pipe dream. The technical difficulties of keeping a conductor down at such extremely low temperatures far outweigh any advantage gained in getting around transmission loss.

What's more, while in principle superconductivity seemed to promise the ability to produce unlimited currents (implying super-powerful magnets), in practice the effect was self-defeating. As soon as any sizeable magnetic field built up, it seemed to interfere with the process, stopping the superconductivity in its tracks. Similarly, for any particular wire there was a current limit, above which the superconducting effect was destroyed. With all these limitations it would be quite a while before practical applications of superconductivity, for all its apparent magic, would emerge. But they certainly would do so. For the moment, though, the disappearance of resistance was little more than a natural oddity. Curious, but of little significance. It is notable that the citation for the Nobel Prize that Onnes received in 1913 does not explicitly mention superconductivity.

Magnetism-free zone

We will come back to the implications of this lack of resistance, but first we need to skip forward to 1933, when the other key quality of a superconductor was discovered by Walther Meissner and his graduate student Robert Ochsenfeld at the German national institute of natural and engineering sciences, the Physikalisch-Technische Bundesanstalt in Berlin. Ever since Michael Faraday's work, physicists had been aware of the concept of a magnetic field, refined and quantified by James Clerk Maxwell. It was the movement of an electrical conductor through a magnetic field that started a current flowing – this was how generators worked. Magnetic fields permeate space and pass through solids (though limited to some degree by permanent magnets). But Meissner and Ochsenfeld discovered what would become known as the Meissner effect. At the exact same temperature as electrical resistance disappeared, a conductor would expel any magnetic field that passed through it, causing the field to bend around the object.

Just as Onnes' original discovery had been unexpected, so Meissner had no idea that this strange magnetic expulsion was going to happen. He and Ochsenfeld were merely studying the variation in the magnetic field in a metal cylinder as it transitioned to being a superconductor. There was no expectation that the cylinder would totally expel the magnetic field. Meissner's discovery proved to be a significant boost for the attempts to explain superconductivity, which had been in the doldrums since Onnes' discovery of the effect. One physicist, Felix Bloch, had commented: 'The only theorem about superconductivity which can be proved is that any theory of superconductivity is refutable.' But the Meissner effect opened up a whole new avenue of exploration.

A quantum chill

These two behaviours, zero resistance and the expulsion of magnetic fields, are the characteristic behaviours of superconductors and between them produce a whole host of practical applications for this eminently quantum phenomenon. However, observing them in action is one thing. Explaining *why* they should happen is entirely another. Onnes had suspected that there was something of the new quantum mechanics involved in superconductivity, though at the time the theory was very sketchy and incomplete. Later on it would be realised that superconductivity could never be explained without invoking quantum effects, far beyond the simple observation that electrons are quantum particles.

It seemed apparent by the 1930s that the electrons in a conductor that carried an electrical current were strongly delocalised throughout the conductor – they weren't particularly associated with one atom any more – but the expectation was that a combination of the impurities in the conductor and interactions between the electrons and the jiggling atoms of the material, constantly in motion even in a solid, would always provide a degree of resistance. This should, as Onnes expected, gradually reduce as the temperature went down – but there was nothing in theory to predict the sudden fall to zero observed when superconductivity kicked in.

The first hint of an explanation came in 1935 when Fritz London, a German scientist who with his brother Heinz had fled Nazi Germany (appropriately enough to work in London), suggested at a meeting of the Royal Society the radical idea that the whole of a superconducting object could act as if it were a single giant atom with the conduction electrons in a fuzz around it, shielding it from the intrusion of a magnetic field and causing the Meissner effect.

At first sight this might not seem much of a step forward. After all, it had long been established that the conduction electrons weren't tightly associated with any specific atom. But the revolutionary, slightly scary aspect of what London was proposing was that this was a quantum mechanical process operating on the scale of a visible, touchable object (at least when wearing thick gloves). Generally speaking, quantum physics, with all its weirdness, applies only at the level of the very small – atoms, photons, electrons and the like. Typically, the biggest thing that can still exhibit quantum effects is about the size of a small virus. Anything bigger and there is simply too much interaction between the particles making it up, causing decoherence. But London was saying that in a superconductor – which could in principle be many metres across – quantum behaviour was the norm.

Quantum vibrations

From the point of view of understanding superconductivity, London's key observation was to suggest that the electrons behaved in a collective manner, sharing a wavefunction, acting coherently, rather like the photons in a laser (see page 129) to produce an irresistible current. But how could electrons, which repel each other and are not exactly sociable, act together in this way? Three key individuals, John Bardeen, Leon Cooper and Robert Schrieffer, would come up with an answer that has been our best understanding of the basic principles of superconductivity ever since. But this wasn't a year or two after London's Royal Society talk. A whole 27 years had passed before things finally came together.

John Bardeen is something of a hero of this book, also involved in the development of the transistor (see page 63). He would receive the Nobel Prize for both pieces of work, putting him in the extremely exclusive club of double winners. Even before the Second World War, Bardeen had been thinking about the relationship between the conduction electrons and the atoms that made up a superconductor. Perhaps, he believed, those atoms could in some way give the electrons a boost.

All atoms vibrate, and in a solid, particularly one with a regular structure, these vibrations produce waves that travel through the material, vibrations that are called phonons. As the similarity of the name to 'photons' suggests, these waves are quantised – they can't vibrate any old how, but are restricted by the structure of the material to vibrate only in certain modes. And though they are true waves, not the 'particles with wave-like probability properties' of a typical quantum particle, this quantisation means that they are still subject to the rules of quantum mechanics.

Bardeen felt that an interaction between phonons in the material and its conduction electrons was significant to superconductivity, and by 1950 it had been shown that the atoms certainly played a part – this wasn't a pure effect of the electrons – because the critical temperature at which a material started to superconduct varied for different isotopes. An isotope is a variant of an atom with different numbers of neutrons in a nucleus, but the same number of protons and electrons. So, for instance, uranium famously has the isotopes U-235 and U-238. These have the same number of protons and electrons (so are chemically identical), but U-238 has three more neutrons in its nucleus than the 143 in U-235.

If superconductivity were purely dependent on electrons then you would expect no difference in critical temperature between different isotopes of the same material, but if, as proved the case, a difference was observed, it suggested that the way the atom as a whole behaved was a part of the superconducting process. What this special vibration seemed to do was to enable a very counterintuitive effect. Electrons are all negatively charged and so we would expect from all our experience with electromagnetism that they would repel each other. But in a superconductor, the electrons have to effectively attract each other to produce the kind of effects that were being seen.

The mattress effect

Physicists have something of a fondness for bowling balls when it comes to providing a visual image to simplify a theory. Where attempts to explain general relativity and its explanation of gravity usually bring in the model of bowling balls distorting a rubber sheet, those trying to explain superconductivity put their bowling balls on a saggy mattress. A first ball is rolled along the mattress, leaving an indentation that is slow to restore, and which a second ball will inevitably follow. The result is that the second ball feels an attraction to the first via the medium of the saggy mattress. In the superconductor, the cooled, slow-reacting ions of the solid's structure act like the limp springs of the mattress, linking the electrons together by giving them a natural path to follow.

The final push to develop a theory to explain superconductivity came in 1957 when John Bardeen, Leon Cooper and Robert Schrieffer came up with what would become imaginatively known as the BCS theory. Leon Cooper had already discovered that pairs of electrons (called Cooper pairs) can be linked together in this mattress-like fashion. Cooper pairs are electrons with opposite spins, pushed together by the slow-responding lattice of the material, acting as if they were a single entity as they pass through a conductor. Once electrons are travelling as Cooper pairs, the pair would have to be broken up if they were to be scattered enough by phonons to produce electrical resistance. Because of the fuzzy location of quantum particles, each electron pair overlaps with others around them, forming a state called a condensate, where all the pairs interact as if they were a single entity. Where it would be easy for a single pair to be broken by phonons, in a condensate, such a break would influence the whole collection of pairs, making it much less likely to happen. The pairs flow like a ghostly whole through the superconductor.

In parallel with developing theories on why superconductors should exist and how they avoided resistance, experimenters were managing to push up the operational temperature of superconducting materials from the highly impractical 1.5 K to a more manageable mid-20s K. We are still talking about very cold materials – this stuff is hovering around –250°C – but such temperatures were more practically achievable outside a specialist lab. Any hope that the gradual increase in working temperatures would continue seemed to be dashed with the development of the basic theory of the mechanism behind superconductivity, as that seemed to suggest that it would never work at higher temperatures than had already been achieved. However, the mid-1980s saw a sudden revolution that (as revolutions often do) threw away the old certainties.

Within months of 30 K being reported as the absolute limit for superconductivity, Paul Chu, Maw Kuen Wu and their team announced at a meeting in New York in March 1987 that they had produced superconductivity at the relatively balmy temperature of 90 K. Rather than a metal, they were using a new ceramic material that combined barium, copper and yttrium with oxygen. When the 90 K breakthrough was revealed it produced a huge upheaval in the field. It had pretty well been assumed that the 30 K limit was the end of the road, and that the potentially very lucrative (as well as fascinating) possibility of a room-temperature superconductor was nothing more than a pipe dream. Now, though, with the 30 K limit smashed, they had made a huge advance towards a temperature that would not require cryogenics to maintain the superconductivity.

The new alchemists

What followed has been likened to alchemy rather than conventional materials science. In the middle ages, before chemistry emerged as a discipline, alchemists would repeatedly try heating and cooling different mixtures. They had no model for what was happening to these elements and compounds as the substances interacted. Instead, they simply tried to reach their goals of transmutation of metals and the elixir of life by haphazardly trying out combinations and seeing what happened. In modern physics, the more likely approach is to try to understand what is happening first, then to enhance the process using that understanding – but after the 1987 meeting, all sorts of mixes of materials, often including rare earth elements like yttrium, were tried just to see how they reacted.

Rare earths, rather paradoxically, are elements that are not particularly scarce; it is just that the minerals in which they were first detected *are* rare. While the use of rare earth elements in the mix had a certain logic – the first superconducting ceramic had been based on barium, copper, lanthanum and oxygen, but this had only reached the 30 K limit, and switching in yttrium seemed to be the key to the sudden leap – the many attempts to try different combinations, in a race to both Nobel and financial success, felt more like the work of those alchemists than a normal piece of scientific work.

There was one guiding principle behind the frantic switching of component elements, though. It had already been observed that putting a material under intense pressure would increase the critical temperature at which superconductivity began. While this wasn't a practical tool in producing high-temperature superconductors, it suggested that finding a way to get the atoms closer in the structure of the material would enhance the interaction between phonons and electrons and could benefit the substance's ability to go 'super' – and one way to do this would be to incorporate bigger atoms in what was otherwise a smaller lattice structure.

Ditching the helium

Within a year, critical temperatures of around 125 K were being reported for materials that substituted thallium, strontium and bismuth for barium or lanthanum in the original mix. There were also occasional flurries of excitement with temperatures reaching all the way up to the 250s K – effectively to room (or at least freezer) temperature. These initial observations were reported once, but then would always fizzle out as they failed to be reproduced. With the 90 to 125 K materials a new plateau had been reached – but it was a very significant one.

Although these were still not superconductors that operated at room temperature, and so were not likely to enable the production of superconducting power lines or superconducting applications in the home, they did have one huge benefit over the traditional superconductors that required temperatures of 30 K and lower. Liquid helium, the essential medium to get down to those temperatures, is expensive to produce. By comparison, liquid nitrogen is more than 100 times cheaper – cheap enough to be widely available in GPs' surgeries, for instance. And liquid nitrogen boils at around 77 K. So these new super-conductors need only liquid nitrogen to function – with the accompanying benefit of significantly less heavy-duty cryogenic storage.

A new mechanism

In parallel with the attempts to produce high-temperature superconducting materials, there was also a race to come up with a theory to explain what seemed to be a very different process to the Cooper pair superconductivity of a basic low-temperature superconducting metal: a race that is still under way. It was realised quite early on that the classic yttrium–barium–copper–oxygen material $(YBa_2Cu_3O_7)$ had a particularly strange structure. Unlike the regular lattice of a metal, this ceramic consisted of a series of stacked copper/oxygen planes, linked by chains of copper and oxygen with barium and yttrium atoms interlaced between the two. This structure produced physical oddities, like a different resistance to electricity at room temperature depending on whether the current was passed along the planes or in a perpendicular direc-tion, passing through them. It is possible that some sort of tunnelling mechanism between the layers could enhance the progress of Cooper pairs.

As yet, though, there is no widely accepted explana-tion as to how these high-temperature superconductors work (the physicists working in this field consider 90 to 125 K to be positively high-temperature). It is still a major area of research for those working to understand the complex structures of these materials and to get a theory that will make it clear why the superconductivity is maintained. The best bet at the moment is that the role played by phonons in a conventional low-temperature superconductor is taken over by fluctuations of the spin – in effect a magnetic mechanism. But this is a long way from being certain.

Ambient supers

While the theorists work away at an explanation, possible observations of room-temperature superconductivity keep cropping up. It may be that these are the superconducting equivalent of the cold fusion furore in the 1980s, when Stanley Pons and Martin Fleischmann claimed to have produced nuclear fusion at room temperatures and pressures, only to find that their experiment could not be reproduced – though it is equally possible that there will be a breakthrough, just as the 30 K barrier was breached.

A good example of the possibilities being explored is the work reported from Tokai University in Japan in 2013. The researchers were using a material known as HOPG – highly oriented pyrolitic graphite. This is like the conventional graphite form of carbon, but with extra bonding between the sheets, giving it unusual behaviour. Pyrolytic carbon is one of the few materials that can levitate above a permanent magnet the way magnets do above superconductors, it is extremely thermally conductive along the plane of the graphite sheets, and it's the most magnetic material at room temperature.

In the Tokai experiment, when two plates of HOPG were dipped into a mix of two organic compounds, heptane and octane, the resistance of the samples dropped below measurable levels. The experimenters also kept a current running without decaying in a ring-shaped container holding the compounds for 50 days. As the researchers put it: 'These results suggest that room temperature superconductor [*sic*] may be obtained by bringing alkanes into contact with a graphite surface.' It is very early days, but such experiments hold out hope for the future.

Metamaterial madness

Other scientists are coming at room-temperature superconductors from a different angle, hoping that the same special materials that make invisibility cloaks possible can also transform the resistance-free world. These are 'metamaterials', where researchers play around with the way substances interact with light or sound or electromagnetism.

Invisibility metamaterials are usually those with a negative refractive index. Refraction is the way light bends as it travels from one medium to another – causing effects like a pencil appearing to bend when put in a glass of water. Negative refractive index means that the light bends in the opposite way to usual, making it capable of bending around something and concealing it – hence the invisibility cloak. But metamaterials have other tricks up their sleeve.

Vera Smolyaninova of Towson University and Igor Smolyaninov of the University of Maryland realised that some metamaterials have a property that makes them very interesting from the point of view of a theory of superconductors first derived by the Russian physicist David

Kirzhnits in 1973. This links the ability of electrons to support superconductivity to a property of a material known as its dielectric response. The lower the dielectric response, the better the electrons interact. In principle, a metamaterial could be made with a negligible or even a negative dielectric response, and this intrigued the Smolyaninovs.

The hope is that by producing a special metamaterial that includes both a metal that acts as a superconductor at low temperatures, like mercury or lead, and impurities of a dielectric material (a substance that is an insulator but that can be polarised to have different charges on opposite ends – they have in mind strontium titanate), it might be possible to encourage the formation of electron pairs at much higher temperatures than is currently possible. It might not reach room temperature (it might not work at all – this is still only theory), but it would be a significant step on the way to designing a material that brought superconductivity to the everyday world.

In the meantime, though, we have had superconductors for just over 100 years and it would be surprising if they had not been used at all, even with the restrictions on keeping them down at cryogenic temperatures. Though superconductors may not have the ubiquity of electronics, they have still started to play a significant role in our lives.

CHAPTER 9

Floating trains and well-chilled SQUIDs

On 10 September 2008, the world held its breath as the Large Hadron Collider (LHC) at CERN was switched on. After some prophets of doom had predicted that it would create miniature black holes or strange matter that could destroy the universe as we know it – while optimists seemed to think we would immediately be led to the much-sought-after Higgs boson – the reality was something of an anti-climax. The world stayed in one piece, and test runs were the order of the day. But nine days after first start-up, things went horribly wrong.

Quenching catastrophe

An electrical fault led to the release of the liquid helium used to cool the massive superconducting magnets that keep the LHC's protons, travelling at near the speed of light, on track. Suddenly the magnets dropped out of the superconducting phase. This shift, known as a 'quench', produced an intense burst of heat as the immense currents were suddenly exposed to resistance, blasting the remaining helium gas out with explosive force. The result was that 50 magnets were damaged and over a year's work would be required to fix it. This was an accident with superconductivity at its heart.

The vast magnets of the LHC are the biggest single application ever made of superconductors, but they represent only a small fraction of the ways that superconductors have come to be used – some of them with much more potential for an impact on everyday lives.

It's true that the early promise was probably overstated. When Onnes first discovered superconductors, his contemporaries had visions of superconducting grids that would carry vast currents around a country with no losses. But the combination of the need to keep the cables at very low temperatures and the way that superconductivity is self-defeating when currents reach a certain level, producing strong enough magnetic fields to destroy the superconductivity, has meant that the applications, while important, have remained rather less than everyday. Unlike transistors or lasers, we don't have superconductors around the house (yet) – they are restricted to specialist applications. Yet those applications are very impressive.

In order to produce the kind of magnets needed by the LHC, or for other applications like the magnetic levitation trains we will see shortly, it was necessary to get around the way that the early superconductors lost their superconductivity in the face of a high-strength magnetic field. Some alloys turned out to have a different kind of superconducting behaviour. These 'type II superconductors' have regions (known as bundles) where the magnetic field is allowed to penetrate the material, and in these regions the material ceases to be a superconductor – but the bundles are surrounded by matter in which the superconductivity persists. The bundles are pinned in place by impurities in the material, preventing them from moving, which would cause a kind of electrical resistance. The result is a superconductor that can cope with the kind of current required for massive industrial-strength magnets.

The quantum magnetic bottle

It might seem that the LHC, as the biggest machine ever made, would also provide the most dramatic challenge for engineers who wanted to make use of superconducting magnets, but the LHC's problems pale into insignificance when set alongside the requirements to build a tokomak. 'Tokomak' is a Russian acronym for a name that roughly means 'toroidal chamber with magnetic coils' (there is some argument over exactly what the original phrase for the acronym was). It is a magnetic confinement reaction vessel for nuclear fusion. In principle, nuclear fusion, the power source of the Sun, would be a superb way to generate energy, far better than any of our existing sources, but it's not for nothing that fusion power stations have been promised for over 60 years and are still a good 40 years into the future. It's a supremely difficult business to contain a small parcel of the Sun on the Earth.

The good news about fusion when compared with conventional nuclear fission power stations is that it uses fuel that is much more easily obtained and that does not produce any high-level nuclear waste. But the problem is that to get fusion operating you have to contain a plasma – a collection of ions – at temperatures of around 150 million°C. This is ten times the hottest temperature in the Sun itself, but the fusion reactor hasn't got the Sun's immense gravitational pressure to help the process along – it needs all that energy simply to make fusion occur. The intensely hot plasma can't be allowed to come into contact with the metal vessel that contains it, or the plasma temperature would instantly collapse and the metal walls would be seriously damaged, so the charged ions have to be held in place by a series of powerful magnets that surround the reaction chamber, which is usually in the form of a ring doughnut (a torus) that has a D-shaped cross section.

Early tokomaks used conventional electromagnets to keep a relatively small amount of plasma in place, but to get the sort of magnetic field strength required for a production tokomak generator requires a whole set of superconducting magnets. ITER, the next-generation tokomak under construction at the moment at Cadarache in the south of France, is not full-scale, but is the last generation of test reactors before a production version is built, and it will have superconducting magnets, as will its successors. Superconductors are a central component of this hugely important step in the production of low-carbon energy.

Just think of the challenges that will be faced by the ITER engineers. Not only have they to create and manage a monstrously hot mass of ions, but it will seem to have a life of its own. A plasma twists and turns as if it were alive in an attempt to escape the magnetic field. Simply keeping the reactor running is a major challenge. But as well as having to produce an immense temperature and keep the plasma in check, the engineers have to manage this writhing inferno right next to superconducting magnets that are cooled within a few degrees of absolute zero. As if it isn't hard enough to make fusion work, the need to keep those magnets cool in the vicinity of such vast temperatures adds yet another challenge to the design.

And yet superconducting magnets will be used, if for no other reason than because it is impossible to envisage how a full-scale generating reactor could develop a strong enough magnetic field without them. They have become a simple essential in the future of electric power stations.

Scanning with a chill

ITER is still in the future, but in one application with a tried and tested value, MRI scanners, superconductors are already in common use. More accurately known as nuclear magnetic resonance, but renamed magnetic resonance imaging because of the negative associations of 'nuclear', MRI is a powerful medical scanning technique that works doubly at the quantum level, relying on a quantum effect to produce powerful magnetic fields and using a second quantum phenomenon to produce its images. The scanner requires extremely strong magnetic fields, most often produced using superconducting magnets, like a small-scale version of the magnetic confinement in ITER.

The scanner works by manipulating the quantum spin of protons in the nucleus of the hydrogen in water molecules, present throughout the body. The subject to be scanned is passed through a magnetic coil with a rapidly varying magnetic field that has a frequency tuned to flip the spin of the protons. When this field is turned off, the protons flip back, producing radio frequency electromagnetic radiation from within the body – in effect, water molecules become tiny transmitters, detected by the receiver coils. Extra, varying magnetic fields are used to pinpoint the signal in three dimensions and produce a cross-sectional image of the body as it passes through the scanner. MRI scans are ideal to distinguish between normal and abnormal tissue, detecting tumours.

The main component of an MRI scanner is its magnet, usually cooled by liquid helium to 4 K (−269°C), producing a superconducting effect. Secondary electromagnets known as gradient coils vary the magnetic field by position to enable a 3D image to be built up. The changes in the field gradient result in rapid expansions and contractions in these coils, producing loud hammering. Without appropriate soundproofing, the coils are as noisy as a jet taking off, at around 120 decibels.

An MRI scanner is the application of superconductivity that is closest to being everyday. Still perhaps something of a rarity, but common enough for most of us to be aware of them. Up to now, the applications of superconductivity, while important, have been limited to research facilities. But the feeling is that now is the time when superconducting applications will really take off around the world. Perhaps the best known such application really involves taking off – if only by a tiny distance – in the mechanism that allows a maglev train to float above the ground.

The levitation train

There is no doubt that railways have the potential to be the best transport mechanism available to us. Unlike flight, rail can make use of low-carbon electricity as a source of power, it is very safe, and is far more efficient and low in pollution than using cars or buses. Because of its separate environment it can also run a lot quicker than any other form of ground transport. High-speed rail now regularly operates at around 250 kph (155 mph), making it comparable on timing to air travel for end-to-end short-haul journeys – because a train can take you straight to your destination with far fewer delays than with airport bureaucracy. But conventional rail is reaching the practical limits to which it can be pushed – and this is where maglev comes in.

Maglev is short for magnetic levitation. We've all played with magnets and felt the almost magical repulsion when the same poles – north to north, or south to south – are brought together. Balance the magnet on a suitable structure and this repulsion enables it to float above a surface. Now add some mechanism for propulsion – typically also based on magnetism – and you have a different kind of train, one that has no friction from its contact with the rails, meaning it can reach higher speeds, and that is far quieter than a traditional railed train. However, keeping many tonnes of train off the ground is beyond the capability of any conventional magnet – and this is where superconductors come in (and one of the reasons why the search for high-temperature superconductors is so important).

There have been a number of experimental maglev trains, and at the time of writing there are two short-run lines in operation, though as yet none is providing a full-scale rail service. Already a maglev train has smashed the world rail speed record: the Japanese experimental MLX–01, using liquid helium to produce superconducting magnets, reached 581 kph (361 mph). The first commercial maglev is now planned in Japan – the Chou Shinkansen, linking Tokyo, Nagoya and Osaka. It's not an overnight development – it could well be 2045 before it's operational – but we can expect speeds of around 500 kph (319 mph). The train is expected to float 10 centimetres above the track, using large helium-based superconducting magnets on board, which both levitate the carriages and provide its propulsion.

Getting maglev right

Although maglev takes away the disadvantages of friction, the train still has to face up to air resistance, which becomes a major issue over 400 kilometres per hour, wasting around 83 per cent of the energy at this speed. And then there's noise. Although the track noise of a conventional train is missing, at this speed the sheer noise of forcing your way through the air has reached an unacceptable limit of 90 dB – the equivalent of a diesel truck around 10 metres away. If even faster speeds are to be achieved (and in theory maglev trains could reach 1,000 kph), the train would have to run through a tunnel that has at least some of the air removed. This might seem an extreme solution, but it has already been suggested as a possible approach for a Swiss metro system – underground lines lend themselves to this 'vacuum train' approach.

We aren't going to see maglev trains taking over our rail networks anytime soon. The noisy political wrangling in the UK in the early 2010s over building a conventional high-speed line (HS2) shows how difficult it can be in some countries to make a major change to infrastructure. But the Japanese have already demonstrated that they are prepared to build dedicated high-speed lines and will no doubt have a similar success with maglev, which will inevitably become more widely accepted over time as air travel becomes less acceptable due to global warming, particularly if room-temperature superconductors could ever be discovered.

Although maglev is potentially a green mode of transport (depending on how the electricity is generated and the coolant produced), it has resulted in one environmental concern, particularly in China, where there are worries about the dangers from 'radiation'. There is often confusion about this, as most people associate 'radiation' with the potentially dangerous output of nuclear reactions – high-energy particles and gamma rays. But electromagnetic radiation from power lines and phone masts – the sources of most concern in our present environment – is just another form of light, in the radio frequency rather than in the form of destructively powerful gamma rays. While it is true that very close exposure to extremely strong magnetic fields can have an impact on the brain, passengers and passers-by do not experience anything close to this with a maglev train and there seems no reason for concern, but it does reflect that such a big change in technology can sometimes be a hard sell.

Although trains have seen the biggest investment in the application of superconductivity to transport, there has been some work done already on a superconducting electric motor for ships. Current, diesel-based ship engines are highly polluting and significant contributors to CO_2 levels (they do also counter global warming because the dirty particulates they emit block sunlight, but this pollution is hardly desirable), but existing electric motors simply can't be scaled up to the size required to power a full-sized ocean-going ship. Prototype motors using magnets based on high-temperature (liquid nitrogen) superconductors have already been tested and could be in commercial use within ten years.

Josephson's quantum genius

It's natural to think of these kinds of applications, because the most familiar uses of superconductors like the LHC's giant magnets and MRI scanners are big machines, undertaking their superconducting on a grand scale – but one of the most flexible uses of this quantum phenomenon comes in at the very small end of the scale in the form of a SQUID. This is not the marine invertebrate, but a Superconducting Quantum Interference Device. To understand these, we first need to meet the Josephson junction, a quantum device that was devised in 1962 by the then graduate student Brian Josephson, who would later win a Nobel Prize for his work.

Josephson is one of the more unusual characters of the physics world. He has been reviled in later life by many of his contemporaries, because he seems to exhibit a wide-eyed acceptance of many phenomena – like the memory of water, and telepathy – that other scientists regard as time-wasting fruitloopery. Yet there is no doubt that in his twenties Josephson was one of the sharpest minds working in physics. A few years ago I went to meet him at Cambridge University, where he had an honorary position at the Department of Applied Mathematics and Theoretical Physics that enabled him to work on his mind–matter unification project.

While he came across as having some of the characteristics that we traditionally associate with a mad scientist,
or at least an eccentric one in the Einstein vein – wild
hair, once he took off his cycle helmet, and a certain
vagueness of conversation – it was interesting that when
I later attended a lecture by quantum entanglement expert
Anton Zeilinger and sat next to Josephson, he was still
treated with great respect by the other physicists present.

Back in the 1960s, though, Josephson was a much more
intense and driven character. He was well known for picking up on lecturers should they make a slip. As one of
them, Philip Anderson, commented: '[Having Josephson
take a course] was a disconcerting experience for a lecturer, I can assure you, because everything had to be right
or he would come up and explain it to me after class.'
Josephson was only 22 when he made the initial discovery that would become the Josephson junction, and he
received the Nobel Prize at 33. What impressed all who
witnessed Josephson at work in those early days was the
completeness of his vision – the way that he had taken
existing theory and built on it to give a total description
of the implications of constructing a Josephson junction.

An expert patent lawyer told Anderson that 'in his
opinion Josephson's paper was so complete that no one
else was ever going to be very successful in patenting any
substantial aspect of the Josephson effect'. And this was
someone who was still two years away from complet-
ing his PhD. From the physicists' viewpoint, Josephson's
work was probably most important because it gave a fundamental insight into the nature of superconductivity,
and in particular it clarified the role of the phase of the

electron pairs in a superconductor. But from the outside this gave new possibilities for practical applications of superconductors.

So what is this effect that brings SQUIDs to life? A Josephson junction consists of a pair of superconductors with a barrier between them, which can be an insulator or a conductor that isn't in a superconducting state. It was already widely understood how quantum particles could tunnel through a barrier (see page 34) – but Josephson predicted that Cooper pairs of electrons could also tunnel through the barrier in some circumstances. Josephson also noted that with an AC current, where the phase varies with time, the junction will act as an incredibly sensitive voltage measurement device, as the frequency will be directly linked to voltage, and frequency is much easier to measure than voltage.

Josephson junctions turn up in a number of applications, from a range of quantum computing mechanisms (see page 230) to very wide-spectrum equivalents of the charge-coupled devices used in digital cameras, which makes them ideal for astronomy applications. But the widest application of this pure quantum effect to date is the SQUID, the superconducting quantum interference device. This makes use of a Josephson junction to detect very small changes in the magnetic field around the SQUID, as even a tiny induced current from the changing field will have a detectable influence on the junction.

At the moment SQUIDs are a little like lasers in the early days of their developments. It was obvious from early on that lasers ought to be useful, but there was a big gap between the speculation and the reality – it took a while for them to settle down and for us to really appreciate what they could do for us. The same is true to some extent for SQUIDs, though they are already starting to make a mark in areas like detecting neural activity from the tiny shifts in magnetic field produced by the brain, and in some types of MRI scanner. One surprising area of development is the detection of unexploded ordnance, known in the trade as UXO (or if you are being formal, 'munitions and explosives of concern'). Frighteningly, between 10 and 15 per cent of bombs and shells fail to explode and are left in the field as a long-term hazard.

Just how long-term the problem is can be understood from the fact that thousands of UXOs from the Second World War are still discovered in Europe every year. Back then, the level of unexploded materiel was well over 25 per cent. Obviously there are also large residues on recent battlefields and on disused military training grounds. In the US alone it is estimated that there is over 40,000 square kilometres of land contaminated by UXOs.

Various methods have been used to detect these unwanted leftovers, from basic metal detectors to complex magnetic field monitors, but nothing can compare with the sensitivity of detecting changes in magnetic field provided by a SQUID in new devices that measure the exact gradient of the Earth's magnetic field below them, allowing them to pick up the location and shape of anomalous objects with unparalleled clarity. Because of the extreme sensitivity, the detector does not need to be as close to the UXO as with a conventional magnetometer, so it is better for coping with heavy undergrowth and water. Using high-temperature superconductors, the equipment can be made sufficiently portable to scan for UXOs both under the ground and under water. At the moment this technology is just being tested, but it could soon be a common sight on old battlefields and testing grounds.

It's likely that we've only scratched the surface of the possible uses of SQUIDs, and like all superconducting devices, they are likely to become much more common should we ever achieve a room-temperature superconductor. But one final application that is worth mentioning is the scanning SQUID microscope. This systematically moves a SQUID over an area to be scanned, using the variation in the magnetic field it detects to build up a picture. This can be used, for example, to scan an integrated circuit to check for short circuits and to ensure that the circuit is acting as expected. The SQUID is not in direct contact with the item being scanned, so the sample can be kept at room temperature and in air, rather than the cryogenic temperatures and low pressure required for the SQUID itself, making the process non-destructive.

Superconducting sewage

Finding superconductors in powerful electronic devices and scanners may not be too much of a surprise, but the last example of an application of superconductivity is a million miles away from the delicacy of SQUIDs. It is in sewage treatment. We live in a paradoxical world that is awash with water – it almost defines our planet – and yet at the same time where there is a shortage of clean drinking water. It shouldn't be that way. The world contains around 200,000,000,000 litres of water for every living person.

If you think of that in terms of consumption, assuming a typical 5 litres a day, the water out there should last over 100 million years. And that would be if it were all used up, whereas we know in practice that most of the water we consume is released back into the environment in short order. Of course that 5 litres represents only our direct consumption. A typical Western water-user will be responsible for up to 10,000 litres a day. In part this is due to washing, watering the garden and flushing the toilet, but also because of the indirect use in the production of the goods we buy and the foods we eat. Just one hamburger takes around 3,000 litres, while a 1 kg jar of coffee requires a massive 20,000 litres. (Though once again, most of this water will be recycled – it doesn't remain in the product.)

The problem, of course, comes not from poor availability of water per se, but the lack of clean drinking water in the right place for those who need it. Arguably this makes any water shortage more of an energy problem than anything else – that's the energy required to clean up the water, whether it is desalination or removing dirt and sewage, and to get the water to where it is needed. And superconductivity can play its part in overcoming this. Most existing waste water treatment – whether cleaning up sewage or cleaning water from a river to use in an industrial plant – is expensive to build and has to be on a large scale to be cost-effective. There are many circumstances where a smaller, distributed system would work better. Surprisingly, superconductors offer a solution to cleaning water that is both more cost-effective and more compact than a conventional treatment plant. What's more, it works more quickly too.

The process makes use of a powerful superconducting magnet to separate off the suspended material in the water. This is obviously fine for magnetic metals, but it seems an unlikely solution for the rest – the typical gunk that we associate with sewage and polluted water. But by adding a substance known as a ferromagnetic adsorbent to the water, this mess becomes accessible to magnetic fields. The suspended particles stick to the adsorbent material, which is then dragged out of the water by the magnets, leaving clean water behind. The only way to get a sufficiently strong magnetic field is to use superconductors.

More to come

These examples are only the beginning of the ways that superconductors could be used in the future. A considerable amount of work has been done on superconducting cables. As we have seen, all the way back to Onnes there has been a realisation that one of the biggest limitations on an electricity grid is its transmission loss – the energy that is lost to resistance in the cables. Although designs exist for superconducting cables than can carry a high current without the magnetic field generated disrupting the superconductivity, they are very expensive compared to a conventional cable and really require something closer to a room-temperature superconductor to be viable.

Still, it remains a significant area of research, as does superconducting energy storage. Clean energy production from wind or solar, for example, has the disadvantage of not necessarily generating the power when it is needed. But storing large amounts of energy for any time until it is required is tricky. Probably the best solution we have at the moment is to use the energy to pump water up a hill to a high-level reservoir, then to use that water to generate hydro-electricity when the demand kicks in. But work is under way, particularly in Japan and Korea, to produce energy storage in the form of a superconducting coil that retains the energy as a magnetic field until required. These are significantly more efficient than water storage, and very fast to store and discharge, but as yet work only on a relatively small scale.

Computing with quanta

Superconductivity may not crop up in your kitchen yet – that would have to wait for true room-temperature superconductors – but it is already playing a significant part in our lives. However, there is another decidedly cool application of quantum physics that could transform an everyday, essential piece of technology. Whether it's using a smartphone to surf the net or get directions in a strange town, or sitting at my desktop writing this book, computers play a big part in my life, as they do for most of us. Even those who wouldn't touch a computer use technology from washing machines to cars and personal video recorders that have computers built in. And being electronic, those computers are quantum devices. But waiting in the wings is the possibility of using computers in a way that puts quantum physics at the very heart of the way they work. Before meeting the computers, though, we have to establish just what put Einstein into a tangle.

CHAPTER 10

Spooky entanglement

Perhaps the most worrying aspect of quantum theory is that Albert Einstein hated it. It's not that Einstein couldn't make mistakes – he could and he did. However, when someone with Einstein's vision fundamentally detests a theory it is not an aversion we should treat lightly.

God doesn't play dice

John Bell, the Northern Irish physicist who played a leading role in the aspect of quantum physics we are about to cover, once commented: 'I felt that Einstein's intellectual superiority over Bohr, in this instance, was enormous; a vast gulf between the man who saw clearly what was needed, and the obscurantist.' Bell, who would make it possible to test whether Einstein was right, came down firmly on Einstein's side.

As we have seen, Einstein had no problem with the quantum nature of light – he was largely responsible for its discovery. But he refused to accept that the quantum events that underlie all of reality are based on randomness and probability. He was sure that if you looked deeply enough you would find 'hidden variables' that gave real, fixed values to the properties of quantum particles, rather than the fuzzy probabilistic view of Bohr, Heisenberg, Schrödinger and friends.

This led Einstein to write in 1926 to his friend Max Born, the man behind the probability interpretation:

'Quantum mechanics is certainly imposing. But an inner voice tells me that it is not yet the real thing. The theory says a lot, but does not really bring us any closer to the secret of the "old one". I, at any rate, am convinced that He is not playing at dice.'

To see what Einstein was getting at, imagine that you toss a coin, it lands on the back of your hand and you cover it without looking at it – a normal coin toss. At this point, assuming it's a fair coin, there is a 50 per cent chance of getting a head and 50 per cent a tail. But we know, in reality, that the coin is showing a particular face. We don't happen to know which one, but the information is there in the system, in a 'hidden variable' – in this case, the coin. According to the quantum brigade, a quantum particle is totally different. Before looking, it genuinely is in a superposition of states. It doesn't have either value, 'heads' or 'tails' (or whatever the possible values are), it just has two probabilities that determine the likelihood of any particular outcome.

Einstein bores Bohr

Although Einstein moaned in writing to Born – and rightly, given Born's responsibility for interpreting Schrödinger's equation as representing probabilities – his main attack was focused on Niels Bohr, the 'ring-leader' of the increasingly mainstream view of quantum theory (the generally accepted interpretation of the theory was even called the 'Copenhagen interpretation' after Bohr's Danish centre of operations). Einstein soon found a way to press his case. Since 1911, the great and the good of physics had met up in conferences known as the Solvay Congresses. These were originally set up by the Belgian industrialist Ernest Solvay with the hope of having an intelligent audience for his own ideas, but Solvay was quietly sidelined, leaving the big guns to get on with a superb scientific conference. At both the 1927 and 1930 Congresses, Einstein buttonholed Bohr and presented to him a series of thought experiments that Einstein hoped demonstrated the failings of quantum theory.

Some of the challenges Bohr was able to dismiss straight away. He commented on one: 'I feel myself in a very difficult position because I don't understand what precisely is the point which Einstein wants to [make]. No doubt it is my fault.' Others he had to work on through the day, or in one disquieting case overnight, when he was able to point out over breakfast, with a satisfying sense of irony, that Einstein was wrong because he had failed to take into account the influence that general relativity predicted gravity would have on the experiment, causing time to run slowly, cancelling out the effect that Einstein hoped would challenge quantum theory.

The EPR paradox

For five years after the 1930 Congress, Einstein was quiet on the topic, and Bohr might have hoped that the attacks had finished. But then in 1935 Einstein published a paper that he believed set the cat among the quantum pigeons, finally and definitively demonstrating a flaw in quantum theory. The irony was that, unknown to Einstein, not only would the effect he described help vindicate quantum physics, it would prove to be a central pillar of modern quantum theory and one that would prove to have dramatic practical applications.

The paper was clumsily titled 'Can Quantum-Mechanical Description of Physical Reality Be Considered Complete?', but it would be universally referred to by the initials of its authors. Einstein was joined by two young physicists, Boris Podolsky and Nathan Rosen, and so the paper became known as EPR.

EPR describes a thought experiment where a particle breaks into two equal parts, which fly off in opposite directions. After a while, according to quantum theory, the particles don't have definitive values for, say, their momentum or position. Instead each merely has a set of probabilities, which collapse into an actual value only when a measurement is taken. The thought experiment imagines making a measurement on one particle when they are far apart. Say we measure its momentum. Then conservation of momentum tells us that the other particle must have exactly the same momentum in the opposite direction. Yet until the measurement was made, these values weren't fixed. So how did the distant particle find out, instantly, what value its momentum should have?

The paper suggested a similar argument could be made for position. It's rather unfortunate that EPR used both these measurements, as it led to some confusion that this was a challenge to the Uncertainty Principle. In fact, the paper considered each measurement separately (and later versions of the thought experiment used a different, often easier to handle property, quantum spin). But the wording was a little misleading. Einstein was not very good at English at this time and relied on his collaborators for the wording. When questioned about why the paper used the two confusing properties, Einstein told Schrödinger: 'Ist mir Wurst', which literally means 'It's sausage to me', an idiomatic term for 'I couldn't care less'.

No more locality

What the authors of EPR concluded was that either quantum theory was incomplete – that there were hidden values that simply weren't known about, rather than true probabilities – or that it wasn't possible to assume that the universe kept things local and real. If quantum theory was right, there had to be what Einstein called 'spooky action at a distance', the ability for distant particles to somehow instantly communicate with each other in apparent contradiction of Einstein's relativity and its assumption that nothing can travel faster than light.

The linkage between these particles is quantum entanglement. It is a phenomenon that we will meet a number of times in the quantum world. In Einstein's eyes this prediction was a counter to the quantum theorists, but Schrödinger turned things on their head when he coined the term entanglement. He said: 'I would not call [entanglement] *one* but rather *the* characteristic trait of quantum mechanics, the one that enforces its entire departure from classical lines of thought. By the interaction the two representatives [the quantum states] have become entangled.'

Initially EPR was seen as little more than an interesting challenge to quantum orthodoxy, but in the 1960s John Bell proposed an indirect mechanism that could distinguish between true entanglement and the work of hidden variables, and by the 1980s, experimenters like the

French physicist Alain Aspect had shown that Einstein was wrong. Entanglement was real. And it would prove a valuable tool in the next generation of quantum devices that are just starting to be conceivable in the 21st century.

The instantaneous communicator

There is something enticing about entanglement. As soon as anyone hears about it, they can immediately see how it could be usefully applied. Make a change to one particle and it is instantly reflected in another at any distance. What you have here is instantaneous communication. We already face problems from communications delays. Phone calls over satellite links can be plagued with unnerving gaps in the conversation. And things are going to get far worse as we take on long-distance space travel. A radio signal from Earth to Mars can take 20 minutes each way, while should we ever make it to the stars, our nearest stellar neighbour, Proxima Centauri, is a good four years away by radio.

Admittedly this isn't something that is likely to be an issue for quite a while, but even those small local delays for communications around the Earth can cause difficulties for electronic systems and voice communications alike. And there is also the intriguing proposition that, technically speaking, an instantaneous communication also makes it possible to send a message backwards in time. By combining the ability to send a message instantly with a receiver where time has run slowly – which according to special relativity only requires the receiver to move at very high speeds – it should be possible to transfer a message to the past.

However, shortly after the realisation that instant communication is a possibility comes the let-down. Although entanglement genuinely does enable information to get from A to B instantly, that information is random and outside our control. Say, for instance, we use the spin property of a pair of particles, which in a particular circumstance might have a 50:50 chance of being up or down when measured. We make a measurement on particle A and it turns out to be down. Instantly, particle B has spin up. But we had no control over which of the superposed values came up when the measurement was made. The fact that B was spin up tells the people at that distant location that A is spin down, but that can't carry any useful information because it merely describes a natural, random occurrence, not a message we wanted to convey.

Despite this absolute limitation, the appeal of the instant communication is so strong that physicists (usually young physicists with something to prove) have often attempted to find a way around it. And some have appeared to get very close. Back in the 1980s, American physicist Nick Herbert was convinced that he had cracked the problem with a design that even Richard Feynman could not find fault with. Herbert tried to make use of the property of photons called polarisation – the property that is used by LCD displays (see page 72).

A polarised message

Polarisation comes in two forms, linear and circular. In the more familiar linear form, the polarisation of the different photons is lined up in the same (or at least similar) directions, while in the circular variety, the polarisation rotates with time, so as the photons travel along, the polarisation direction corkscrews around the direction of motion. Herbert's idea was to start with an entangled pair of photons, then send one, the 'local' photon, through a polarising filter, choosing to make it either linearly polarised or circularly polarised. These two forms of polarisation would correspond to 0 or 1, allowing the system to instantly communicate in binary, as the polarisation of the second, 'distant' photon would instantly reflect the local one, however far away it was.

The problem with this approach is that, given a single photon, you can't ask, 'Is it linearly or circularly polarised?' It is possible, for instance, to check if it is polarised in a particular direction and get a 'yes' or 'no' answer, but not to make the required distinction. So Herbert planned that before the polarisation is applied to the near photon, the distant photon would be sent through a laser gain tube, a device that produces multiple copies of the photon. Then the distant beam of photons would be split in two, half going through a linear polarisation detector and half through a circular polarisation detector. Herbert reasoned that this would make it possible to detect the outcome.

Unfortunately there was a fault in his logic. Laser gain tubes don't make perfect copies of photons. They produce multiple photons that are similar (for instance in energy) to the original, but they aren't identical in terms of their quantum properties. In fact the so-called 'no cloning theorem' proves that it is not possible to create a second photon that is absolutely identical to another. This complex but thorough quantum physics theorem shows that the closest that can be achieved is to make a perfect copy while destroying the original in a process we will soon meet in more detail, called quantum teleportation. But this doesn't make it possible to 'breed' multiple copies. The laser gain tube loses the whole point of the exercise and the beams produced do not make it possible to deduce what had happened to the single, near photon.

Quantum secrecy

No one since Herbert has even come close to dreaming up a means of using entanglement for communication, and it is highly unlikely that anyone ever will. However, the randomness that gets in the way of sending an instant message has proved a strength for a fruitful potential use of entanglement – in encrypting data. Keeping messages secret is a challenge that has been faced ever since we began to communicate. In the early days, concealment was the most common approach. A message might be written on a tablet that was then covered with wax, on which an innocent cover message was written. There were even secret notes written on messengers' shaved heads that were then concealed as the hair grew – not exactly ideal for our modern, high-speed world.

As early as Roman times, it became obvious that something more flexible was needed, a way of communicating openly in which eavesdroppers could not understand the message. The most obvious way was simply to use a different language, which could be effective as long as that language was unknown to any eavesdroppers. As recently as the Second World War, Navajo 'code talkers' were used to send US military messages in the Pacific theatre on the reasonable assumption that the Japanese enemy had little chance of interpreting the language. Failing that, an artificial language could be constructed, and this is effectively what a code is, where a word (which can be nonsense or plain English) stands in for a different meaning.

Thanks to this total lack of connection with the meaning, codes provide a very strong way to conceal a message, but they are also inflexible – if you don't have a code for a particular word or message, you can't send it. Codes also depend on having a code book at each end of the communication line, which is inevitably susceptible to being copied, allowing the messages to be read by a third party. For these reasons, the vast majority of coded messages are actually ciphers, mechanisms that provide a rule to systematically change the letter values in a message.

The simplest cipher is the so-called 'Caesar cipher', used since Roman times, where the letters to be sent are simply shifted along the alphabet by a fixed quantity. So, for instance, if that quantity is 3 (apparently popular with the Romans), A becomes D, B becomes E, and so on. This way, a message like 'start a bombardment' becomes 'vwduw d erpedugphqw', or even better, 'vwduw derpe dugph qw', with the characters split into regular blocks so there are no clues to be gained from word length.

A simple cipher like this is relatively easily broken, especially once you know that some letters appear more frequently than others in any particular language. With a frequency table it is possible to guess which substitutions have been made and gradually decipher the message. Today it would be very unusual to use anything other than a key-based cipher. Here, rather than having a single shift value, each letter in the message is displaced through the alphabet by a different amount. A simple key might be just a word that is repeatedly 'added' to the text (adding on the position value of each letter of the key in turn). Others use complex mechanisms to produce that key.

The unbreakable message

In principle, most key-based ciphers can be broken, as happened with those generated by the German Enigma machines during the Second World War, because it is usually possible (if very difficult) to duplicate the mechanism used to provide the key. But one type of cipher that has been around since the early years of the 20th century is totally unbreakable. This is called a 'one time pad'. The idea is that you have an entirely random key – a random set of numbers that are added to the text to encrypt it, then taken away to decrypt. The key is as long as the message, so after encryption you have a random set of letters with no way to be able to guess how the encryption was done. Even if you managed to break a small part of the message, it gives no clue on how to break the rest.

There is a downside to the approach, though, which is similar to the problem of safely deploying codes. To use a one time pad, those random values have to be available to both sender and receiver of the message. And however the values are transmitted – whether by radio or more securely as a physical object, printed on paper or stored on a memory stick – they are vulnerable, liable to be intercepted. And once an eavesdropper has a copy of the random values, any message using this key is wide open to being read.

This is where quantum physics, and entanglement in particular, come in. Because there is true randomness at the heart of the behaviour of quantum particles, it is easy enough to use this to generate the one time pad just before use – but there is still the issue of getting the information from one end to the other, so the receiver can decrypt what the sender has encrypted. Quantum entanglement has a way around this.

By basing the key on measurements made of entangled values, the key literally does not exist *at either end of the communication* until the moment it is used. It is never stored, never transmitted. Imagine, for instance, we are using the quantum property of spin, which will be either up or down when measured, and allocate, say, 'up' as 1 and 'down' as 0. We now have a source for a key in binary form. With the right entangled particles, the spin has a 50:50 chance of being up or down when measured. So up until the sender makes that measurement of a sequence of particles, the key does not exist, but the moment she does, the receiver has the appropriate value to decrypt the message.

The only potential flaw to this approach is that a third party could intercept the entangled particles immediately before the key is used, producing the key, noting it and passing it on. However, this can be detected, as it is possible to check if a particle is still entangled or not, and interception would break the entanglement. If this check is undertaken every few particles, the link can be kept secure without a risk of interception.

Beam me up

Another impressive use of quantum entanglement is to bring to life one of the most dramatic aspects of the *Star Trek* franchise – the transporter. This supposedly scans the particles in something or someone to be transmitted from the starship *Enterprise* to the surface of a planet and reproduces them at the destination. It was done in the TV show primarily to save money on expensive model work showing shuttles landing, but it makes for a remarkable concept – at least as far as science fiction goes. The reality is full of holes.

One problem was identified, in rather an unsubtle way, in the 1950s movie *The Fly*, originally starring Vincent Price and remade in 1986 with Jeff Goldblum in the role of the unfortunate scientist who invents a matter transmitter. A fly gets into the transmission chamber along with the test subject and the result is a horrible human/fly hybrid. While the specifics of the movie are very unlikely – there would be far more (if less grotesque) problems with all the air molecules in the chamber – the underlying concept that it would be very difficult to safely scan and reproduce all the atoms in a person is more than true. In part this is because there are so many atoms in a human – around 7,000 trillion trillion – that it would take thousands of years to scan them all at any conceivable rate.

Send in the clones

There was one other impossibility for the matter transmitter to face. As we've already seen, the 'no cloning' rule means we can't make a copy of a quantum particle. And make no mistake, such a transporter is not really moving something from A to B: it is making a copy of the original from A at B. However, quantum entanglement stops this from being a problem. By using a pair of entangled particles we can transfer a particle's quantum state to another particle. This gets around the no cloning rule because the information is never revealed. It gets from one particle to the other without being measured.

Effectively what happens is that an entangled particle at the transmitter is interacted with the particle to be transmitted. Depending on the outcome of that interaction, some information is sent conventionally to the receiver and that information tells the receiver what to do to the second entangled particle to produce a final particle in the same state as the original. As part of the process the original particle has its quantum state scrambled – and anything made of these particles would be disintegrated.

That would give any scientist who invented a matter transmitter like that in *The Fly* pause for thought. What they have really is not a matter transmitter but a matter duplicator that produces an exact copy at the remote location while destroying the original. If you passed through such a transmitter, as far as everyone was concerned the person at the other end would be you. It would have your memories, your thoughts, be perfect in every physical detail. But the 'you' that stepped into the transmitter would be dust.

In practice, because of the near-impossibility (quantum physicists learn never to say 'never') of scanning a whole person and then recreating them at the other end, the quantum teleportation process is limited to individual quantum particles. But this has proved a boon for a special type of device that needs the ability to transfer quantum information from one place to another without ever finding out what it is – an essential because taking a measurement will change the particle irretrievably. These are devices that make use of the quantum peculiarities of superposition and entanglement to do far more than any normal computer could achieve in a lifetime.

The next big step of the quantum age will be the introduction of the quantum computer.

CHAPTER 11

From bit to qubit

Computers have become central to our everyday lives. We are educated by them, work on them, entertain ourselves with them and use smartphones to communicate with each other. In Victorian times 'computers' already existed – but were nothing more than people who computed, individuals who worked with numbers. The first possibility that the term could mean something different came with the work of Charles Babbage. In 1821, the 30-year-old Babbage was laboriously checking a new set of astronomical tables.

Computing by steam

Babbage was a rich man who certainly didn't need such mind-numbing employment to earn money – his work on the tables was a favour for his friend, John Herschel (son of the astronomer William Herschel, the discoverer of Uranus). Babbage was clearly not impressed by the mind-numbing job and is said to have cried out: 'My God, Herschel! How I wish these calculations could be executed by steam.' Whether or not, as legend has it, this was the inspiration for his masterwork, soon afterwards Babbage set out to design a mechanical calculator that could replace such tedious repeated calculations effortlessly and without error. It would be a computer extraordinaire.

His Difference Engine was a geared mechanism where the input values were first set up on a series of dials and then a handle was repeatedly turned to literally crank out the results. Babbage only ever built a small part of his Engine (though a full-size working version has been constructed by the Science Museum in London), as his mind was already on greater things. The problem with the Difference Engine, compared with a human computer, is that the machine was limited to undertaking the simple arithmetic operations that were fixed by its gearing. Brains enabled human computers to be much more flexible, and Babbage wanted to avoid having the instructions hard-coded in the mechanism. He turned away from the Difference Engine. The government was not amused. They had given him £17,000 – the equivalent of between £1.2 million and £13 million today, based on simple cash value or the labour value of the money respectively. Babbage's political masters expected something in return, but he only ever gave them a part-finished machine.

The notion that distracted Babbage from what no doubt would have been a very useful machine was the idea of using a series of flexible instructions, fed into his mechanical computer using cards with patterns of holes punched into them. Such cards were already in use in the French Jacquard loom, which had revolutionised the silk weaving industry by specifying patterns as a series of holes in a chain of fabric-linked cards. In Babbage's mind, the same mechanism that controlled the loom could also control a computing engine, providing both the data and the instructions on what to do with that data. This machine could, in principle, undertake practically any calculation.

The Analytical Engine, Babbage's great vision for a true mechanical computer, was never built. It was probably impossible to do so faced with the engineering limitations of the day; but Babbage did think through the requirements, developing, for instance, the idea of having separate parts of the machine to act as a memory and for processing the data. He could have got further, at least on paper, if he had taken more notice of the work of the woman who is often described as the first computer programmer, a slightly odd portrayal given that she never saw a computer.

The woman in question was Augusta Ada King, Countess of Lovelace. Born Ada Byron, daughter of the Romantic poet, she had known Babbage since her teens, and might well have married him had her mother not set her heart on marrying off her daughter to nobility. Ada King (or Ada Lovelace as she tends to be known) translated a paper on the Analytical Engine written originally in French by the Italian scientist Luigi Menabrea. Rather than just work on the document, she appended long notes on how the Engine might be used. It would be exaggerating to say, as is often claimed, that she wrote programs, but she certainly gave thought to how she would go about it, and probably would have done so had not Babbage coldly turned down her offers of help. As it was, the Analytical Engine never saw the light of day, and nor did the opportunity to program it.

The true father of computing

Babbage is often, if inaccurately, called the father of computing, which is arguably the fault of Winston Churchill. Thanks to Churchill's paranoia we have only relatively recently become aware of just how much the West owed to a young man called Alan Turing. Babbage could at best be called the grandfather of computing, as Turing was indubitably its father. Babbage's work was a dead end in the evolution of computing, with no real technical link to the computing industry that would eventually develop, but Turing set the entire direction it would take. Thanks to Churchill's insistence that Turing's work on code-breaking at Bletchley Park during the Second World War be heavily suppressed – primarily in the hope that the UK's post-war enemies would not realise how much code-breaking was achieved – it has taken a long time for all the pieces to be put in place that show just how much we owe to Turing.

Alan Turing has become something of a mythical character. He certainly proved brilliant as a code cracker and again in his analysis of the nature of computing, laying the foundation for all modern digital computers in a way that still isn't always recognised. Part of the reason for his mythical status is the often-repeated suggestion, started at his inquest, that his death at the age of 41 was suicide, as a result of the persecution he received as a homosexual when this was still illegal in the UK.

It is certainly true that Turing suffered a cruel and unnecessary punishment for his 'crime' by undergoing a session of 'chemical castration' – treatment with female hormones, which was offered as an alternative to prison. (Turing was pardoned by the UK government in 2013.) It has been suggested that this treatment so devastated Turing that he committed suicide by eating an apple laced with cyanide. In reality, all the evidence was that Turing had recovered fully from the treatment and was happy when he died. He had a laboratory next to his bedroom in which he had been running a chemical experiment that is thought to have produced the cyanide fumes that killed him – to all appearances an unfortunate accident, rather than suicide. But however he died, there can be no doubting his legacy.

The ultimate limit

Turing showed the power of computers in his work at Bletchley and explained their workings with his development of a hypothetical 'universal computer' – but he was also the first to realise their limitations. We are used to computer manufacturers bringing out more powerful, faster computers every year. The growth in computing power of a processor has been described for many years by 'Moore's Law', an observation made by one of the founders of Intel, Gordon Moore, back in 1965 that the number of transistors on a chip (a rough measure of the computer's power) seemed to double every year. This was later modified to doubling every two years. And ever since then, Moore's prediction has come close to reality. But this unchecked growth can't go on for ever.

At some point we will hit physical limitations, where the increasing miniaturisation of the circuits on a chip mean that it is reduced to dealing with individual quantum particles and there is no further to go. What's more, down at that level, quantum effects like tunnelling can cause serious problems for the operation of the chip. But even more significantly, Alan Turing showed that there is a fundamental limit to the capability of software, whatever the hardware happens to be – hence his universal computer design. Turing showed that there are some problems that can never be solved by a computer and many more that would take longer than the lifetime of the universe to crack. For example, it proved impossible to write a program that would decide if any other program, fed into it as input, would come to a stop or keep running for ever. A familiar example of a problem that would take far too long to calculate is working out the best route from A to B on the road network. When a satnav computer does this, it has to use approximation, unable to reach a definite conclusion.

Turing's imaginary universal computer consisted of a paper tape on which zeroes and ones could be written, read or erased, but it is easier to think of a conventional electronic computer as a series of switches, each with two possible positions, represented by 0 and 1. All that any computer does, from Turing's conceptual version to your laptop or smartphone, is spend its time flipping these switches between 1 and 0, under the control of a series of instructions that are themselves stored as ones and zeroes. On a computer chip holding memory, each tiny component part of the circuit is the equivalent of a pair of electronic devices – a capacitor and a transistor. The capacitor is the memory itself, with an electrical charge representing the bit, and the transistor acts as a switch to enable the state of the capacitor to be read or changed.

Quanta to the rescue

Although, as we have seen, any type of electronics is a quantum device, the limitations envisaged by Turing face a new challenge in the form of a truly quantum computer, which takes things to the next level. Instead of being based on the 0 or 1 value of an individual bit, stored as a whole collection of electrons making up an electrical charge, a quantum computer uses the state of a single quantum particle, known as a qubit (pronounced cue-bit), to hold its value. Because this is a quantum particle, it can be in a superposition of states. Typically, for instance, you might use the spin property of a particle to store information. As this can be in a superposed state with probabilities of having either value when measured, instead of having a bit that can be either 0 or 1, we get an expansion of the capability of the computer.

At its simplest this becomes obvious as you add extra bits to the computer. Imagine a three-bit traditional memory. This can store a single number between 0 and 7. But three qubits can store all eight numbers simultaneously in superposition. What's more, in a sense a qubit can store an infinitely long decimal. This is because a superposition doesn't have to be 50:50. If you imagine the probabilities of being up and down corresponding to the direction of an arrow between up (100% up) and down (100% down), then with the right relationship of probabilities, the direction of that arrow could be used to specify any number between 0 and 1. A qubit is, in effect, analogue rather than digital, holding a smoothly varying quantity rather than an incremental value. It gives a computer greatly expanded capabilities. At least in theory.

That proviso comes in because dealing with qubits is not easy. Superpositions tend to collapse, and it is difficult to make use of the values, because observing a quantum particle changes its state. But manage to get it right and a quantum computer promises to be able to perform operations that would take a conventional computer the lifetime of the universe to undertake. And, as we shall see, we even have some of the programs that will perform otherwise impossible tasks already written, at least in concept.

It's easy to see why working with quantum computers is so fiddly if you compare what happens if you check the value of a traditional bit with reading a qubit. A traditional bit, depending on the charge it holds, will either come up 0 or 1. And unless you change its value it will continue to do so indefinitely. Simple and straightforward. But imagine a qubit that is stored as the spin of a quantum particle. Remember, this is a property of the particle that won't have an actual value until measured, just a collection of probabilities. The quantum computer might treat the value as 0.4, because the qubit has a 40 per cent chance of being spin up and a 60 per cent chance of being spin down. But if we, outside the computer, take a measurement of the spin, we will still only get either up (0, say) or down (1). We will get 0 forty times out of a hundred and 1 sixty times out of a hundred. But we can't tell that by simply taking a single measurement. The qubit works in analogue but reads out in digital, which can be hugely frustrating.

Vapourware algorithms

There is a striking parallel between those thinking about the ways we could program a quantum computer and Ada King pondering on how to use Babbage's non-existent Analytical Engine. Just as King was able to consider the kind of processes the Engine would use, so some scientists have thought about how a quantum program would run. Even though we don't yet have large and stable enough quantum computers to do much more than was achieved in a breakthrough in 2001, when a quantum computer calculated the factors of 15 correctly as 5 and 3, there are already algorithms – essentially logical representations of programs – that, given a working computer, would enable us to crack some otherwise insoluble problems.

The approach that was used to work out the factors of 15 is based on a hugely powerful quantum computing algorithm produced by Peter Shor at AT&T. Dating back all the way to 1994, it is a way for a quantum computer to work out the factors, the numbers that are multiplied together to produce a larger number. It can do this with numbers that would take the best conventional computers thousands of years to work out. Strangely enough, although this is potentially very useful and is one of the simpler quantum computing algorithms, it is something that most computer scientists had hoped would never be made possible.

The experts are concerned because the difficulty of working out these factors lies behind the encryption used to keep our transactions on the internet secure. Whenever you see a little padlock symbol in your web browser – typically, for instance, when you are entering credit card details – it is using an encryption mechanism called RSA. This is a so-called public key/private key system. It is a method of encoding data where there are two keys – the values used conceal the data. One, the public key, is used to encrypt the message. This can be given out to anyone. But without the private key it is impossible to decrypt the message. And this is kept by, say, the bank. This means anyone can encrypt the message they send to the bank, but only the bank can read the result.

This mechanism was devised by three computer scientists at MIT in 1977 (Ronald Rivest, Adi Shamir and Leonard Adleman, hence RSA). They weren't the first. The approach was first developed by the British scientist Clifford Cocks in 1974, but Cocks was working at the UK government's intelligence communications centre GCHQ, and his discovery was kept secret until after RSA was released. (Shades of Churchill's dead hand.) The private key is based on two very large prime numbers, which are multiplied together to produce the even vaster number that forms part of the public key. Without knowing the two prime numbers it is impossible to decode the message. Finding out the prime factors of such a huge number would take far too long using a conventional computer, so the mechanism remains secure. But Shor's algorithm breaks through this problem and makes it possible to produce the factors with a relatively modest quantum computer – hence the concern of the computing community.

In search of a quantum needle

Breaking secure websites isn't, of course, the only use of Shor's algorithm – it can be applied to a wide range of computing problems – but that nagging ability makes it of serious concern. Another powerful way to program quantum computers has already been established in Grover's search algorithm. This too has been around a while, dating back to 1996. What Lov Grover realised was that a quantum computer had a huge advantage over a conventional one when it came to sifting through a database without an index – so much so that the paper Grover contributed to the staid journal *Physics Review Letters* on the subject was entitled 'Quantum Mechanics Helps in Searching for a Needle in a Haystack'.

Although computers are very fast at finding information, they use clever techniques to make this happen. The simplest of these is an index. So, for instance, if you imagine a database of customers for a company, it will use indexes to make it possible to get to the information quickly by, say, name or customer number. But it may well not have an index of their driving licence numbers (assuming your database included these) – so if there were a million people in the database, you might have to look through 999,999 of them to find the correct entry. That would be unlucky – but on average you'd expect to have to look at half a million records. With a quantum computer, the Grover search algorithm could guarantee finding the entry using just 1,000 searches because its quantum algorithm works with the square root of the number of entries.

It might seem that there's no need for this, as Google and other search engines seem capable of searching vast quantities of data in the blink of an eye – but they manage this by constantly building indexes, a process that consumes increasingly vast amounts of energy and storage. With such a huge quantity of information, a true quantum search engine could transform the search industry. What's more, because it works on probabilities, rather than looking at each individual item, the Grover algorithm can do more human-like searching with vague, fuzzy criteria.

Grover gives the example of finding someone's phone number. Maybe it's someone you met the other day. You can remember that his first name is John, and that he has a common surname, but not exactly what that surname was. Say you think he's a Smith with a 50 per cent probability, Jones with 30 per cent and Miller with 20 per cent. You also remember that he said he could see the Tower of London from his flat, and that the last three digits of his phone number are the same as the last digits of your doctor's number. With just these sorts of fuzzy information, a typical starting point when attacking unstructured real-world requirements, Grover's new algorithm enables a quantum computer to home in on the right result vastly quicker than anything possible with a conventional search.

First catch your computer

These two applications – fast search, and factoring large numbers – are between them the current big hopes for quantum computing, though there may well be many other uses for it that have yet to be devised. According to Lov Grover: 'Not everyone agrees with this, but I believe there are many more quantum algorithms waiting to be discovered.' As yet, the development of these algorithms remains significantly more advanced than the computers they could run on. Although, as we will discover later, there is a commercial computer now available that is called a quantum device, it isn't capable of making use of these full-scale quantum computer algorithms.

There are dozens of teams around the world working on quantum computers, and some have got a handful of qubits briefly operating, but these are still very much breadboard university rigs, often involving cryogenic cooling – certainly not the kind of thing you can buy down at PC World. Just as Babbage struggled to overcome the basic engineering principles that held back his computing engines, because his designs pushed tolerances to the limits, so real-world quantum computers come up against the tight tolerances of quantum mechanics, always in danger of collapse of superposition and loss of entanglement.

Inside the engine

To understand the difficulties quantum computer engineers face, we need to discover a little more about how computers actually work. We are used to interacting with a pretty graphical interface, but way underneath that, the processor is churning away, manipulating whole series of zeroes and ones. There are only really three processes that make everything from a word processor to a super-realistic video game possible. The value in a bit might be read, it might be copied or it might be changed. Those processes are undertaken using devices called gates, which can choose whether or not to change a bit dependent on the value of another bit.

Gates are simple devices that are the physical representation of a rule. In principle there are all sorts of ways they can embody a rule. Typically in the computers we use they are electronic, with transistors providing the control, but they could be mechanical – I had a working model of a computer when I was young that worked on a set of physical gates – or the gates could be represented by a set of actions on Turing's paper tape. We have already come across gates in electronic form when looking at the development of the transistor (see page 65).

Quantum computers also have gates, but they tend to be significantly more complex than their traditional equivalents. In part this is because of that fact of life in the quantum world, the 'no cloning theorem'. The good news is that quantum entanglement gives a partial workaround. The process of quantum teleportation (see page 222) means that it *is* possible to generate an exact copy of a quantum particle if the state of the original particle is lost in the process.

Entanglement makes copying possible by transferring the properties to the new particle without ever finding out what they are. But even so, quantum gates have a much trickier job than conventional ones. So, for instance, the nearest that the quantum world has to a NOT gate is an X gate. The NOT gate we met before simply turns 0 into 1 and 1 into 0. The X gate takes a qubit with a probability A of having the value 0 and a probability B of having the value 1 and produces a qubit with a probability B of having the value 0 and a probability A of having the value 1. After passing through the gate, the probabilities are swapped.

Juggling quanta

Much of the logic required to operate quantum computers, and the structure of the underlying gates, has been established since the last century – the difficult bit is getting the computer to work. The problem is not in the software but the hardware, and specifically with handling quantum particles. Say, for instance, you wanted to use photons as your qubits. It's easy enough to make photons. Turn on a traditional light bulb and you are pumping out around 100 billion billion photons a second – plenty to be going on with. But that isn't particularly helpful if you want to work on a single particle.

It isn't impossible to perform computations with a whole batch of quantum particles – in fact this is how the first actual demonstration of the Shor algorithm worked. This particular experiment, undertaken in 2001 at IBM's Almaden Research Center by Isaac Chuang and his team, made use of molecules rather than photons. Rather than attempt to handle a single molecule, they took a billion billion of their bespoke seven-atom fluorine/carbon molecules, each of which effectively acted as a 7-qubit device. They used radio frequency pulses to influence the nuclear spins of the atoms and an NMR (nuclear magnetic resonance) device, like an MRI scanner, to detect the outcome.

The clever part was that these were specially designed molecules, structured so that the atoms within them would influence neighbouring atoms. So, for instance, one atom might undergo a flip of spin direction only if the next atom along was already in the same state, making them act like gates. Having large quantities of the atoms meant both that it was possible to detect the results with the NMR device and that the inevitable errors would be drowned out by the majority of correct values. The experimental computer, as we have seen, managed to deduce that the factors of 15 were 3 and 5 – not exactly challenging, but demonstrating Shor's algorithm in practice.

However, this kind of approach, using the atoms within a molecule as qubits and averaging across a cloud of molecules, works only with a small number of qubits, so it is still likely that a serious working quantum computer would have to resort to handling individual quantum particles, and most of the work now going on in this field is based on this assumption. There are mechanisms to produce single photons, while atoms (or at least the charged versions that have gained or lost electrons, ions) have been manipulated individually since the 1980s.

Pinning down the particles

Photons have some significant advantages as qubits because they are easy to make and are stable, and this means that the gates required to set up a quantum computer are often as simple as a combination of beam splitters (see page 115). At the end of the calculation it should be easy to read off the results as well – photon detection technology is highly advanced. But the downside of using photons as your qubits is that they don't easily interact with each other and they won't keep still, although they can be trapped in a small space using a mirrored container, or slowed down using materials like Bose–Einstein condensates.

Atoms, or more likely the charged version, ions, make more tractable qubits that stay in place and interact easily, though the gate mechanisms become more complex than they are for photons. Ion traps are among the most popular approaches used by quantum computing experimenters. Others are looking at the possibility of computing with SQUIDs (see page 200). Perhaps the most obvious approach, though, is using quantum dots, solid-state traps that can hold a single electron. This approach is less messy than many experimental quantum computing rigs – more like the everyday environment of a commercial computer – but like every other quantum computer, a dot-based device faces the problem of decoherence.

Isolated from the world

To make a quantum computer work, the qubits have to be isolated from the world around them, interacting only with gates and other qubits. In practice it is almost impossible to isolate a quantum particle entirely and it soon interacts with the matter around it, producing decoherence, the process by which it loses its superposed state and ceases to have any value in the computer. This is doubly a problem with entangled particles. As Xiasong Ma, working in the entanglement capital of the world, the University of Vienna, commented: 'Entanglement is hard to prepare, hard to maintain and hard to manipulate.'

Stretching the time before decoherence occurs – in early quantum computers this was measured in millionths of a second – is an essential if the devices are to be used for any real-world task. Quantum computers may be fast, but they are liable to be employed on heavy-duty tasks that take more than a few milliseconds to compute. One possible approach to overcoming decoherence is the 'hot potato' mechanism, where any particular qubit is used only for a very short space of time before its properties are passed on to a different quantum particle using teleportation. But the more a computer is scaled up, the bigger the decoherence problems become.

Keeping those qubits fresh seems to be something that we are getting better at, though. In 2013, a team working at Simon Fraser University in Canada managed to keep a collection of qubits in a superposed state for around three hours at temperatures near to absolute zero or 35 minutes at room temperature (though the process still had to be started with the qubits cooled to cryogenic levels). This is around 100 times longer than had ever been achieved before. The qubits were formed from the spins of phosphorus ions held in a chip of extremely pure silicon. However, before getting too excited about this development it is worth noting that it involved around 10 billion ions, all in the same spin state, which wasn't modified – which makes it useless as part of a real quantum computer. It isn't a trivial step to go from this to having the ions in different states and able to interact without decoherence.

A long road

There is something to be said for the thoughts of Richard Hughes of the US National Laboratory at Los Alamos, who describes the efforts to date as 'still in the vacuum tube era of quantum computing'. As we have seen, in the early days of traditional electronic computing, computers were built using these tubes or valves. It simply would not have been possible to have scaled up the 18,000-valve ENIAC to have the same capabilities as the multi-million-transistor chips we use today. The answer wasn't to construct a better valve, it was to develop a whole new technology that does the same job but in a very different way, and the same may be the case in the race to scale up the quantum computer to a stable, workable number of qubits.

One positive example of the kind of sideways step necessary to bring quantum computers into the practical world was made in 2013. The entanglement process, which enables the qubits to interact and provide the scaling that makes quantum computing so powerful, was shifted from complex experimental rigs using difficult-to-handle quantum particles to a solid-state chip. A team at the University of Queensland were the first to entangle two locations on a chip. Rather than atoms, electrons or photons, the experiment used 0.2mm aluminium structures – massive in quantum terms – as artificial atoms that could act as qubits and that were entangled, linked via a superconducting microwave waveguide, a metal tube that acts on microwaves like a fibre-optic cable for visible light.

The harmony of discord

Most researchers believe that it is only a matter of time before the bugs are ironed out, but avoiding decoherence and keeping qubits entangled presents such difficulties that a few have suggested it won't ever be possible to create a useful quantum computer this way. However, hope has arrived from a strange direction. It seems that it may be possible to make use of the messiness of a collection of less-than-pristine qubits and still do computation, employing something known in the trade as discord. The possibilities first emerged in 2001 when it turned out that the apparently successful quantum computer that discovered the factors of 15 using Shor's quantum algorithm was actually flawed. It couldn't have been safe from decoherence. The quantum computer was operating at room temperature, and the entanglement between the computer's seven qubits would have collapsed long before a result could be obtained. Yet it worked.

In a traditional quantum computer, a quantum gate might take two or more pristine, entangled qubits as input and the result would be read off afterwards. But it was discovered that putting through the gate one traditional, cleanly set up qubit, carefully protected from interaction with its environment, and one qubit in a more 'normal' messy state that has been subject to measurement, would also enable a computation to take place. The qubits couldn't be entangled, but there seemed to be enough of an interaction to allow the quantum calculation to proceed. 'Discord' is a measure of how much a system is influenced by observing it. A traditional classical system has zero discord, but any quantum system in a superposed or entangled state has a positive discord, and it seems that discord can give a degree of correlation, a kind of pseudo-entanglement that isn't so susceptible to collapse, linking together a mix of pure qubits and messy ones.

The output of a discordant computer is a little unnerving to those used to the precision of traditional computing. Because of the messiness involved, the results are not exact but have to be averaged across a number of runs – but if this is done, the result seems to be reliable. What we have in a discordant computer is still a quantum device. It does require at least one pure qubit that is protected from decoherence, and though the rest of the qubits are in normal classical states, discord itself is a quantum effect. The whole process collapses and fails to work if that one pure qubit is allowed to go messy. But up until then, it is almost as if the addition of noise and disorder in the messiness of the rest of the qubits makes for a better, more stable quantum computer than one that is carefully protected from its environment – a decidedly hopeful thought for anyone attempting to build a commercial model and struggling with the menace of decoherence.

At the moment the approach is of limited use, because we only have the maths to be able to make use of very simple set-ups with discord linkages. As yet, the experimental physicists are waiting for the theoreticians to catch up. But there is a lot of promise there, and discord is being taken seriously. 2012 saw the first discord conference in Singapore, with over 70 researchers attending. This hybrid approach to quantum computing certainly has legs. Which is probably a good point to introduce D-Wave.

D-Waving but not drowning

Back in 2007, the Canadian company D-Wave Systems unveiled what it claimed to be a quantum computer. Not everyone agrees. This might seem crazy – either something is or isn't using quantum effects. In reality, there is no doubt that the D-Wave computer is quantum technology, but what is uncertain is whether the particular approach taken, which is very different from conventional quantum computing, will give the kind of scaling benefits that make all the effort worthwhile. D-Wave is an 'adiabatic quantum computer'. This means that rather than have the sort of quantum logic gates we have discussed, you have a collection of qubits set up to use a process called quantum annealing.

Quantum annealing means that the qubits will try to reach their lowest energy state, and it requires the calculation to be set up in such a way that the lowest energy state should deliver the answer required. It's a bit like searching for the lowest point in a landscape by first taking an overview that finds the lowest collection of points and then homes in on the lowest point within that area. Quantum tunnelling allows a kind of random walk through peaks of energy that would act as barriers to find lower and lower energy states until the minimum is discovered. There is a danger with any algorithm for finding a minimum in this way that it could settle on a point that isn't the actual absolute minimum, which is something those writing algorithms for adiabatic quantum computers have to be aware of, but in principle it does provide a mechanism for finding a solution.

An experimental version of such a computer achieved an early triumph back in 2006 by smashing the 'factoring 15' trick by providing a solution to the factors of 143, using only four qubits. This isn't using the speedy Shor algorithm, but rather a special 'adiabatic' algorithm that encourages the system to naturally produce the factors of the number. And while it is clear that Shor is much, much faster than any conventional means of producing factors, it has not been demonstrated that the adiabatic algorithm is any quicker. The claim has been made that a D-Wave computer can solve some problems 'as much as 3,600 times faster than particular software packages running on digital computers'. This is true, but it involved comparing a specially tuned algorithm designed for a specific purpose running on a $10 million D-Wave with off-the-shelf software running on a PC. This isn't clear proof of an advantage.

The latest version of D-Wave with 503 qubits, bought by Google and installed at NASA's Ames Research Center, is certainly a lot more sophisticated than the early test machine – and it looks more like a commercial computer (though think more of a supercomputer than a laptop) – but there are still similar doubts about whether the adiabatic process will deliver real quantum computing benefits. To date, the best success from the D-Wave approach has been in producing an algorithm to recognise images. This is obviously valuable to a search engine like Google, but again there is not yet good evidence that D-Wave can achieve this faster than an equivalent conventional machine. The jury is out, but one thing is certain: D-Wave is not a general-purpose quantum computer, but rather a very specialist device that can run only certain limited algorithms.

It's good to talk

Getting quantum computers to work is one thing – achieving a quantum version of the internet is another. Being able to establish entanglement between two distant points is already of value, because it would make it easy to use entanglement-based quantum encryption (see page 220) between those locations. For this reason, a considerable amount of work has been done on achieving an entanglement link across the equivalent distance of a link between an Earth station and a satellite, initially beaming entangled photons from building to building in Vienna. (This was harder than it sounds: the owners of one of the office blocks had to be kind enough to agree to temporarily replace their glass after it was discovered that the special heat-reflecting windows totally ruined the entangled signal.) After some preliminary test work using an experiment on the International Space Station, the first experimental satellite that will beam entangled particles to two well-separated locations on the Earth – the essential for entangled quantum key distribution this way – was launched by the Japanese in 2014.

There is also the challenge of achieving a similar distribution of entangled particles through optical cables, either for encryption or so that quantum computers can be linked at the qubit level and share computations. A particularly sparkling idea for a way to make this happen was published in 2013 from work at Delft University of Technology in the Netherlands. The novel concept was to use diamonds.

In the experiment, entangled qubits were held in two diamonds 3 metres apart from each other, though obviously this was intended as a proof of concept for a larger-scale separation. The reason this is of interest, when remote entanglement has already been demonstrated in more conventional qubits like trapped ions, is that it is thought that connecting many qubits in diamonds may well be much easier than taking other systems up in scale, because the structure of the diamond crystal was discovered to have a particularly valuable behaviour.

The qubits in the Dutch diamonds are based on impurities in the crystal. A pure diamond is a perfect lattice of carbon atoms, but by combining nitrogen atom impurities with gaps in the lattice, a qubit can be formed based on the spin state of electrons held in a gap. As yet this is an inefficient process, producing only one entanglement in 10 million attempts, but it is expected that this can be made significantly more efficient. The big advantage, though, that diamond has over ion traps is that the qubits are relatively stable at room temperature. By contrast, ion traps have to be heavily cooled. The entangled diamonds can get away with the higher temperature because the crystal lattice shields the qubits in its midst from potential sources of decoherence. What's more, it has already been demonstrated that qubits, which inevitably decay, can be passed around nearby gaps in a diamond, keeping them stable for seconds rather than the microseconds typical of a qubit – and diamond has the potential to be more scaleable than many other qubit technologies. It's early days, but diamond is definitely one to watch, and it may end up being a quantum computer scientist's best friend.

Although it isn't essential, most quantum computing at the moment relies on the use of cryogenic environments to keep the qubits relatively stable. Being supercool isn't just about electrical and magnetic effects, though. It can transform the properties of fluids too.

CHAPTER 12

It's alive!

Imagine having a ring of liquid sitting on a table. You give it a quick stir to get it rotating. And it goes on. And on. For ever. Just as currents will flow indefinitely in a superconductor, a superfluid experiences no viscosity or friction. It has no internal stickiness; there is nothing to stop it moving. Fill a cup with superfluid liquid helium and this bizarre substance will climb up the walls of the cup of its own accord, roll over the edge and drip out.

A special liquid

Just as Heike Kamerlingh Onnes was the first to observe superconductivity, his achievements in the supercool also meant that he first observed a superfluid in action – though he did not discover the strange behaviours mentioned above. As helium was being cooled towards the critical temperature, the tiny observation windows in his reaction vessel enabled Onnes to see the liquid boiling away. In fact the cooling process made use of this boiling, as it involved removing the helium vapour from immediately above the liquid, taking away the fastest-moving atoms and leaving the slower, cooler ones behind.

As the temperature fell below 2.17 K, the appearance of the liquid was transformed. The seething of the boiling process abruptly stopped, leaving a calm, level surface. Clearly some kind of phase change had occurred. What happened (though Onnes did not realise this) was that some of the liquid helium had become a superfluid, with no viscosity and no thermal resistance. Bubbles form in a boiling liquid because it is not entirely uniform. There are hotspots where more of the liquid turns to vapour, forming a bubble. But the superfluid was such a good conductor that it transferred any build-up of heat away before a bubble could form, leaving a calm, untroubled surface.

Onnes noted this without taking it any further, while an understanding of what was happening had to wait until the 1930s, when Pyotr Kapitsa in Russia and John Allen with Don Misener in Cambridge discovered the full reality of superfluidity. Kapitsa had begun his low-temperature work at the Mond Laboratory in Cambridge, where he developed a new mechanism for producing large quantities of liquid helium shortly before returning to Russia (with some of his equipment) in 1934 to set up the Institute for Physical Problems. Here, in 1937, he would discover superfluidity in collaboration with the other pair back at the Mond.

Escapologist bosons

Superfluids are escape artists. Given the slightest aperture, their lack of viscosity means that the natural movement of molecules is enough to enable them to work their way out over time. In effect, despite being liquids, they act as if they were gases. More dramatically, a superfluid held in an open container like a cup will form a thin film up the surface, over the lip and down the outside, meaning that over time it will drain out of the cup, apparently defying gravity. The effect is caused by a very similar phenomenon to superconductivity. In a superconductor, electrons pair up and produce overlapping wavefunctions through the material. In a superfluid, larger bosons get a linked wavefunction to form a type of substance known as a Bose–Einstein condensate.

It's worth making a brief detour to explain this 'boson' word, as it has frequently been misused in the media (typically when referring to another boson, the Higgs boson). A boson is a type of particle named after the Indian physicist Satyendra Bose, so it is pronounced *boze-on*, not *bosun*, as many newsreaders mangled it. All quantum particles are either bosons or fermions, referring to the kind of mathematics that describes their behaviour. Bosons follow Bose–Einstein statistics while fermions behave according to Fermi–Dirac statistics. The two types of particles have different spin values – half-integer values for fermions and integer values for bosons. In behaviour, bosons are happy to collect together in large quantities in exactly the same state, while no two fermions can be in the same quantum state in the same system, with a combined wavefunction. This is known as the Pauli exclusion principle and is one of the fundamental rules of quantum mechanics.

For basic particles, 'matter' particles tend to be fermions and the force-carrying particles (like photons) are bosons. But combined particles, like atoms, can be either, depending how the spin values add up. Superconductivity and superfluidity are dependent on the particles behaving as bosons, which is why it is necessary for the electrons in a superconductor to pair up – their half-integer spins combine to make an integer spin: the pair becomes a boson. We can see the significance of this with the two types of helium.

'Standard' helium – helium-4 – has two protons and two neutrons in its nucleus. This even number makes it a boson, so it is capable of getting into the special state where it becomes a superfluid relatively easily. But the isotope of helium with just one neutron – helium-3 – is a fermion with an odd number of particles. It is only when helium-3 atoms pair up to form a boson that it becomes a superfluid, which occurs at a significantly lower temperature.

More than an oddity

For a long time it seemed that superfluids were nothing more than an oddity that made for impressive lab demonstrations, but recently they have been used in specialist gyroscopes to measure very precise variations in gravity and as a 'quantum solvent' that makes other matter clump together of its own accord, a phenomenon that is very useful in spectroscopy. Spectroscopy makes use of the interaction between matter and light to discover the composition of a material, and is often performed on a gas. The peculiar physical nature of the quantum solvent means that dissolved particles act as if they were gaseous, so spectroscopic analysis can be performed on materials that are hard to study in a gaseous form.

The way that a superfluid acts as if it were a gas can also be used to provide a kind of super-refrigerator. Fridges rely on a physical effect where a liquid forced through a narrow aperture naturally causes cooling, because it flash-evaporates, taking energy out of the liquid and reducing its temperature. Superfluids are particularly effective at this process and were used, for example, in

the Infrared Astronomical Satellite, an infrared space tel-
escope launched in 1983, which employed supercooled
helium in such a refrigerator-like system to keep its systems extremely cold, avoiding vibration.

But one very special application of a superfluid stands
out above all others – because this is a substance that
can capture the most elusive natural phenomenon. Some
superfluids can trap light.

Slow glass

Imagine a special window, which it took light a year to
travel through. Set it up in front of a beautiful view for a
year – then you could place it in a house anywhere and
you would have that view for the next twelve months.
You wouldn't be looking at a TV picture of what was
out there, but the real view, simply seeing the light as it
came through the window twelve months later. It's science fiction (Bob Shaw wrote an excellent book called
Other Times, Other Eyes based on this concept of 'slow
glass') – but it is also strangely close to the truth of a
special kind of superfluid, a way to use a Bose–Einstein
condensate to influence the speed of light itself.

We are used to the speed of light being referred to as
a universal constant. Nothing, we are told, can travel
faster than light. And this is true – but it is also an unac-
knowledged shorthand. What we really should say is
that nothing can travel faster than the speed of light
in a vacuum. This is 299,792,458 metres per second
(exactly and definitively because the metre is defined as
1/299792458th of the distance light travels in a second),
and it is indeed a universal limit. (Forgetting the oddity of

quantum entanglement.) But send light through a medium like air or glass and it slows down. It's easiest to see why at the quantum level of individual photons.

We tend to think of a transparent substance letting light pass through unaffected, but in fact those photons are irresistibly drawn to interact with electrons, boosting the electron energy and disappearing. Soon after, the electron will re-emit a photon and it continues in this constant dance of QED through the material. But that process inevitably slows the photon down. In glass, for example, light travels at about two-thirds of its speed in a vacuum. This slowing, incidentally, leads to the strange blue glow that comes out of the liquid that surrounds some kinds of nuclear reactor. The reactor is spitting out electrons that zap through the water at a higher speed than the speed of light in water, and the result is a kind of optical sonic boom, called Čerenkov radiation.

So in principle we could use a piece of glass to make that special window. The only problem is that two-thirds of the speed of light in a vacuum is still very fast. A piece of glass holding a year's worth of light would be 5,000,000,000,000 kilometres thick. But a superfluid can do a whole lot better. At the end of the 1990s, a Danish scientist working in the Rowland Institute for Science at Harvard University, Lene Verstergaad Hau, used a Bose–Einstein condensate to slow light to a walking pace.

Inside the condensate

In the Harvard experiment, sodium atoms were cooled until they formed a Bose–Einstein condensate. Usually a condensate would be opaque, but Hau's team used a laser to blast a path through the material. This laser modified the condensate so that photons in a second laser beam, following in its path, became entangled with the particles of the condensate, resulting in these photons passing through the material at a much-reduced speed. The early results got the speed of light down to around 17 metres per second, but with further work it was reduced to well under a metre per second. However, this was not the final trick that this superfluid had up its sleeve.

In 2001, Hau's team were experimenting with the effect of reducing the intensity of the initial beam that produces the path, called the coupling laser. If the beam strength is gradually reduced to nothing, something remarkable happens to the second beam. It never emerges. It's not that it has simply been absorbed, though, like light entering a black cavity. Because when the coupling laser is restarted, the second beam flows out of the condensate. The material manages to trap light inside it, producing a kind of mix of matter and light known as a 'dark state'.

The dark state would later be developed in a way that could be very valuable in the future of quantum computers. In 2006, Hau's team managed to get a photon-based qubit to be absorbed into a condensate, then released when later required, leaving the qubit unaltered. What seems to be happening is that the qubit is transferred to a wave in the condensate, which then recreates the optical qubit. Because the condensate has a single wavefunction, the sodium atoms behave coherently and don't lose the quantum information. In the experiment, the qubit was passed to a second condensate a very short distance away (160 micrometres) and was recovered from that second, totally separate condensate after the qubit passed from one to the other in the form of a physical wave in the matter.

A jumble of photons

This is all very remarkable, though it isn't truly slow glass. The problem is that light travelling in different directions would take different times to get through the glass. In fact this is always the case, but usually the effect is so small that it can be ignored. Imagine, though, you had a piece of glass a centimetre thick that light took a year to go through. A diagonal path through the glass might take 1.4 years to get through. So the image seen on the other side of the window would be a jumble of photons that didn't make any sense. What would be needed to make true slow glass would be a combination of the slow transit time and some kind of holographic technique that would allow the whole image at the outer surface to pass through as a single unit.

However, this limitation would not be a problem if you only wanted light to pass through, rather than an image. Imagine, for instance, you hang a plate of slow glass with a 'time depth' of about twelve hours a few metres above a road. During the day, the light would gradually pass through the glass. As it got dark, that light would start to pour out, illuminating everything below, not with artificial light but with actual sunlight, just a little delayed in arriving. It wouldn't matter that the view was scrambled, because you aren't trying to look through a window, just making use of the light it emits. It would be a perfect artificial light that didn't even need electrical power to keep it going.

In practice this is still science fiction. Hau's Bose–Einstein condensate required expensive cooling near to absolute zero – far more expensive than running any street lamp – and worked only with laser light. But the principle of a slow glass that it demonstrates is still a fascinating opportunity to think of possible applications for the future, should such a room-temperature material ever be made.

Feel the force

In 2013, the interaction between a Bose–Einstein condensate and light once again entered the realms of science fiction, but this time a remarkable new physical behaviour was likened to a lightsaber. The press had a field day: '*Star Wars* lightsabers finally invented'; 'Scientists Finally Invent Real, Working Lightsabers'; 'MIT, Harvard scientists accidentally create real-life lightsaber'. To be fair, one of the scientists behind the discovery, Professor Mikhail Lukin of Harvard, fuelled the fire by commenting, 'It's not an inapt analogy to compare this to light sabers' – and Harvard no doubt appreciated all the publicity – but the reality was very different from the headline frenzy.

In some ways a better comparison, had the media been as familiar with quantum physics as you now are, would have been to say that this was an optical equivalent of a Cooper pair (see page 183), the linked electrons that make superconductivity work, because Lukin and MIT professor Vladan Vuletic had produced light 'molecules', pairs of photons that were linked together. Not a lightsaber in sight, though I admit it doesn't make such exciting PR.

The pairing of photons is a big deal, not because you could use it to make a lightsaber (you couldn't even come close), but because generally speaking one of the most frustrating aspects of trying to use light in computing, particularly quantum computers, is that photons are solitary creatures. They ignore each other, passing through other photons as if they weren't there. In everyday life, this is a good thing. Just think for a moment about what's going on in the air right in front of your nose. It is threaded with billions of photons heading in all directions. There's the visible light that is enabling you to see (and to read these words). Radio frequency light being used by radio, TV, phones, Wi-Fi, Bluetooth and more. Infrared from your radiators. It's a sea of inter-threading light.

Now imagine that these photons did interact. All hell would break loose, optically speaking. In the massive collection of collisions between all the photons you would lose the ability to see, and all our radio-based technology would no longer work. Not a great picture. But in a controlled way, in a quantum computer, or indeed any computer using photons rather than electrons, it would be very valuable if we could get photons to interact with each other, because we don't want qubits to be totally isolated. There has to be some interaction to enable the computer to work.

Switching light

There has already been one step made in this direction. Also in 2013, an MIT/Harvard/Vienna University of Technology collaboration with many of the same participants as the 'lightsaber' experiment produced a 'light transistor', a switch that allows a single photon to decide whether light is transmitted or reflected by the device. It consisted of a pair of mirrors with a gas of supercool caesium atoms between them. The mirrors were carefully positioned to create a resonant cavity that produced a quantum effect meaning that light could pass straight through the mirrors; but fire a single photon into the caesium gas and the quantum state changed sufficiently to prevent almost all of the light from passing through.

This is exciting for those working on quantum computers because it is just a single photon, with all the quantum abilities of superposition of states, that is causing the effect. What we have here is a kind of real, small-scale Schrödinger's cat where the quantum peculiarities of the photon are transferred to the light beam it is controlling. Although the initial experiment was only a proof of concept – a real computer would need a solid-state equivalent of the supercooled gas – it is a valuable contribution to the concept of optical computing. But probably not as significant as the optical molecule.

Manning the Rydberg blockades

So what was going on in that 'lightsaber' experiment? When one or more electrons in an atom have been pushed up to very high energy levels, the excited atom is known as a Rydberg atom, and it influences nearby atoms, preventing them from becoming excited to the same state. If two photons are pumped into the medium used in the experiment, the first sets up one of these 'Rydberg blockades' and the second is held up until the first photon has moved further through the medium – the two photons become linked together, pushing and pulling at each other as they sequentially interact with a series of excited atoms.

The experiment is at the very early stages of practicality, but the experimenters suggest that it may lead to the ability to create more complex structures out of photons – perhaps even crystals of light – and that it should be possible to use these structures to create the sort of logic gates necessary to make computers work, but employing photons rather than electrons. Funnily enough, it's hard to imagine anything further away from the blistering, fixed-length light beams of a lightsaber than these delicate photonic molecules – yet the excitement the press release generated is perhaps not so unrealistic, as this genuinely has real potential to enable us to do much more at the quantum level with light.

When we deal with such optical computing, or strange states of matter like a Bose–Einstein condensate, or the fascinating phenomena of superconductivity, it seems that we are working at the extremes, seeing something we would never find outside the lab. But that's a very parochial view. As we have already seen, quantum physics is constantly escaping from the world of science into our everyday existence.

CHAPTER 13

A quantum universe

It's almost pointless to try to imagine a world without quantum technology. Taken at the base level, everything depends on atoms and light and their quantum dance of interaction. But even if you limit your view to those technologies like electronics that are designed around quantum physics rather than naturally making use of it, our modern lives would fall apart without it.

Quantum tech

As I sit typing on my iMac, a device filled with quantum technology, viewing this text on a quantum LCD screen, I can see phones (mobile and wired), a bank PIN device, a laser printer, a CD drive, even a lightsaber (though I have to admit, that's a toy). I simply couldn't do my job any more without quantum support. And all the time those quantum developments that are currently on the leading edge are moving outwards towards being practical, every-day applications.

Take quantum encryption. This is not just a matter for ivory-tower scientists – or even the academics who venture out into the world, like Anton Zeilinger with his experimental link-ups in Vienna. The fact that a wide range of companies are working on quantum key systems, from big, established computer businesses like IBM and Toshiba to specialists like the American MagiQ and the Swiss ID Quantique, shows that there is a real potential for making practical use of this powerful quantum effect.

Entanglement in space

The essential to transform the distribution of quantum keys from a laboratory one-off to a widely available service that can work reliably for everyday use is to set up a kind of quantum version of the internet – or at least a very singular private internet. To make this practical would require a network of stations that could distribute quantum keys to remote locations so that, for instance, a bank in the UK could safely communicate with a bank in the US. The ideal location for these quantum stations is in space, as a satellite would have a wide reach of possible targets. This is why Anton Zeilinger's team in Vienna spent time sending entangled photons from building to building over distances comparable to that to a satellite. The same channels could also be used to set up a kind of distributed quantum computer, linking qubits in one location with qubits in another using entanglement.

Keeping entanglement live across a distance of many kilometres isn't an easy process. Usually a radio or laser signal sent over a long distance is composed of huge numbers of photons. Many of the individual photons will be lost along the way, scattered by air molecules or otherwise interacting with matter. But a quantum signal works at the level of individual photons. Losses along the way can seriously disrupt the process. It's not possible to have multiple copies of a single bit of the key, or someone could intercept part of the stream without necessarily disrupting the entanglement. This makes satellites as switching stations doubly attractive, as a fair amount of the distance the photon travels will be in space where there is little matter with which the photons can interact.

There are a number of attempts under way to get an experimental satellite quantum station in operation, from Anton Zeilinger's work with the European Space Agency to set up an experiment on the International Space Station (which has the advantage while still experimenting of having astronauts on hand to tweak the process) to a collaboration between the Canadian Institute for Quantum Computing and the Canadian Space Agency with the Japanese satellite mentioned earlier (see page 248). Although it is a way off, it seems likely that there will eventually be a commercially available, satellite-based 'quantum overlay' to the internet that will allow easy, totally unbreakable communication between two locations on Earth that could be up to half a globe apart (depending on how high the satellite orbits).

A moral vacuum

Like many new technologies, this possibility is both useful and worrying. It means that in principle financial transactions could be made totally secure, and this process may arrive just in time to prevent the current public key/private key system from being devastated by the use of Shor's algorithm in quantum computers to factor those large primes. Our everyday online secure transactions – the ones with the little padlocks that appear in the web browser – could become protected by quantum key distribution from the next generation of these satellites. But at the same time, this truly unbreakable communication would potentially be a boon to terrorists and spies, who could communicate securely without fear of government interception. A system like RSA can be set up in such a way that security services can gain access to encrypted files. This may raise civil liberty hackles, but sometimes it may be justified. With good quantum key distribution there is no back door. Secure is secure, whatever the message.

Like any new technology, the applications of quantum physics are morally neutral. They can provide certain capabilities, and it is up to us to decide whether we want to make use of those capabilities. But the genie of knowledge is out of the bottle and is not going back.

In some ways quantum physics and its applications has a parallel in the way that electricity was treated by scientists and engineers in an earlier age. They had no idea what electricity was, but they could make use of it to deliver the increasing benefits of the electrical age. We understand the kind of electrical effects that delighted the Victorians far better now, but we have hit up against a far bigger barrier when trying to comprehend the quantum world because we don't have a mechanism for understanding. In one sense, physics can never deliver the absolute truth of the nature of reality at the quantum level. All it can ever do is provide us with models. Just consider comments on what things are like at the quantum level from two great Nobel Prize-winning physicists, Richard Feynman and Steve Weinberg.

What are quanta?

Feynman, as we have already seen, said: 'I want to emphasize that light comes in this form – particles.' Feynman happened to have been talking about light, but he would have said the same for other quantum particles. Weinberg, on the other hand, has written: '[T]he inhabitants of the universe [are] conceived to be a set of fields – an electron field, a proton field, an electromagnetic field – and particles [are] reduced to mere epiphenomena.'

It's easy to think that the older Feynman made his comments earlier, and his ideas were replaced by Weinberg's new thinking, but Feynman's particle definition appeared in a book published in 1985, nearly ten years *after* Weinberg made his comments. It seems perhaps more natural for non-physicists to go with Feynman's particle view, because on the whole, stuff as we experience it does come in particles, not in the form of a field. Fields seem a cerebral concept that clearly can't be *real*. It seems wastefully excessive to require a field that fills the entire universe to define the location of a single electron, rather than have an apparently simple tiny object. The majority of modern physicists, however, would say that Weinberg was correct. In practice they are both wrong – or both right, depending on your points of view.

The fact is that both ways of describing quantum phenomena fit the observed results. And we have no other way to distinguish between them. We have no way to look at an electron, say, and see what it 'really' is. All we can do is test the models against the observations we *can* make – and there are lots of these – and see how well those models hold up, how well their predictions match what is observed. Both the particle and the field approach fit equally well. The field approach happens to work more easily with the maths that is currently in use, which is why modern physicists prefer it – but Weinberg was entirely wrong to refer to particles as 'mere epiphenomena'. They are just as real (and just as imaginary) as fields.

Guessing in the dark

Bruce Gregory in his book *Inventing Reality* points out that the job of the scientist, and the physicist in particular, is a bit like looking at a complex clock that is totally sealed up and trying to decide what mechanism inside is producing the results we observe on the outside. We can come up with various different theories as to why it displays various indications on its dials at any particular time. There could be a clockwork mechanism inside, there could be a radio link, picking up an externally broadcast time signal – or many other possibilities. It could be, for instance (though this is probably unlikely anywhere other than the fictional Discworld), that there is an imp inside, turning a handle to match the regular beat of its own pulse. Similarly, the way that information reaches the dials could work through a whole host of different mechanisms.

All we can do, bearing in mind that the clock can never be unsealed, is to see how well our theories do at predicting what will turn up on the dials. We can match our theories against observation. And those theories that match best and that are useful at predicting what will happen in the future are the theories we can go with. Sometimes there will be several theories, all of which produce the same results and that it may be possible to show mathematically are equivalent. If that is the case, it is totally arbitrary which of the theories we use. We might prefer one over another because the mathematics is easier

– or because it sits better with our natural inclination – but it is only a matter of personal preference about what language we use to describe what is happening.

Most importantly, we need to get away from the idea that our models *are* reality. When we describe how an atom behaves or what happens when light interacts with matter, we are not describing reality. Instead we talk about our model that happens to fit the observed results rather well. Think of that sealed clock once again. It is entirely possible (though I hope unlikely) that the clock works because there really is a very small impish person inside it pulling levers. That could be reality. But provided that person acts in such a consistent way that it *could be* clockwork inside, and such that my maths for clockwork predicts what the hands will do, it doesn't really matter, except at a philosophical level. Science is more pragmatic than philosophy. It has to be.

And that's just as well. If our scientists had spent all their time trying to get to the heart of reality we would never have been able to make use of quantum physics. It would be a sterile discipline with no significance to the outside world. But in fact physicists have given us models that fit reality* well enough that we can make use of all the strangeness of quantum behaviour.

Building with quanta

Even at the most fundamental level we are just beginning to open up the possibilities of quantum physics. It is possible to use our quantum expertise to produce new

_____

* Whatever reality is – and we are never going to be able to open the metaphorical clock.

materials that behave quite unlike anything known to nature. Scientists wince when bad science fiction comes up with 'some new element previously unknown to Man' that is inevitably superstrong or otherwise miraculous, because we know a lot about the elements and what possible gaps there are in the periodic table – but the field is much more open for new materials.

We are used to the three basic forms of matter – gas, liquid and solid. Physicists usually recognise five, adding plasma – matter that is heated so much that it loses or gains electrons and becomes a collection of ions – and the Bose–Einstein condensate. But as Malte Grosche, head of the Quantum Matter Group at the Cavendish Laboratory in Cambridge points out, there's an interesting parallel with chemistry.

Chemists have only around 100 elements to play with. But when it comes to compounds – different ways to combine the elements – the possibilities are endless, from simple two-atom forms like sodium chloride to the magnificent structure of the vast DNA molecules that form our chromosomes. Similarly, by using quantum methodology, it is possible to make new states of matter in which the way the electrons self-organise can transform the nature of a material.

It is early days and very difficult to predict how the exotic-sounding list of examples Grosche produces will change things. He speaks of 'unusual particle–hole condensates, such as spin or charge Pomeranchuk order', and of 'skyrmion lattices in chiral magnets, magnetic monopoles in spin ice materials, and topological insulators'. They may sound like items in the equipment cupboard of a badly written science fiction starship, but they are real, and they offer new building blocks for quantum devices and materials we are yet to imagine.

Seeing quantum effects

We also see increasing examples of quantum effects cropping up in 'macro' objects – things we can see and interact with. The best-known are phenomena such as superconductivity that we have already met, but perhaps the strangest example of all emerged in an experiment undertaken at Paris Diderot University in 2005. Yves Couder and Emmanuel Fort had put a bath of oil onto a platform that produced vertical vibrations in the mass of the oil. They then dripped small drops of oil (but big enough to be visible with the naked eye) onto the surface of the bath. It might sound like the sort of experiment that students dream up after a night on the town, but the results were startling.

The obvious expectation was that the drops would disperse into the body of the oil, helped on their way by the vibration, but in fact they remained whole, bouncing up and down on the surface and sending waves rippling out across the oil. As the experimenters turned up the power, the droplets started to skip across the waves, bouncing back and changing direction just before they reached the edge of the bath. Couder and Fort called the animated drops 'walkers' – their behaviour was almost as if they were alive – but also noted a distinct similarity with the quantum world. The bouncing droplets and the waves that propelled them had a kind of unity. It was a sort of visible wave/particle duality.

Not surprisingly, the pair began to experiment further with their 'walkers' and found that these weren't their only similarities with the quantum world. They managed to duplicate a version of the quantum Young's slits experiment where a set of single particles would build up an interference pattern as the drop passing through one slit interacted with waves that had passed through both. And they observed a droplet version of the kind of quantised orbits that had set Niels Bohr off on the path to quantum theory in the first place.

In particular, the walkers reflect the early ideas of Louis de Broglie, who thought that wave/particle duality emerged from actual particles that were accompanied by a 'pilot wave' that brought in the wave-like behaviour that is observed. In many ways, the oil drop in the Paris experiments is steered by the pilot waves on the surface of the oil bath. A few physicists believe that there is a direct connection with quantum physics – that this is a real quantum effect – while most see it as an elegant and entertaining parallel to the processes we believe underlie quantum effects, not directly based on the same phenomena but providing an excellent analogue model for what happens.

There are certainly some clear differences between this type of quantum phenomenon and the usual version. Quantum particles don't 'run down' – without an interaction they carry on for ever, whereas the waves in the oil bath have to be constantly fed with energy from the vibrator. And the waves in the Paris experiment are limited to two dimensions, whereas each quantum particle has probability waves in its own set of three dimensions (so two particles require six dimensions, and so on). This independent set of dimensions seems essential for entanglement to function, so the walkers are unlikely ever to exhibit that most quintessential of quantum phenomena.

Falling in love with your model

The peculiarities of quantum physics, and the distinction between models and realities, is easy stuff to misunderstand, even for the professionals. In fact a lot of scientists find it difficult to remember that their models aren't real. In his book *The Grand Design*, co-authored with Leonard Mlodinow, Stephen Hawking proclaimed that philosophy is now dead because science can address the questions that philosophy was once required for. Yet this portrays a naive idea of the nature of science. In fact, scientists would have a much better understanding of what they actually do if they were forced to study a touch of philosophy of science along the way.

Take, for instance, Brian Cox, who we previously met allowing quantum particles to be in more than one place at a time. Cox does a great job of popularising science: I use him only as an example to show how easy it is to get it wrong when talking about models and quantum theory. In one of his books (written with Jeff Forshaw), Cox writes: 'Quantum theory provides a description of Nature that, as we shall discover, has immense predictive and exploratory powers ...' As we have seen, quantum theory is certainly immensely valuable in helping us understand what Nature will do, and to enable us to harness that in everything from electronics to lasers. But quantum theory doesn't *describe* Nature. It can't. All it can ever do is predict the observations that we will make of Nature, and that is a very different thing.

Avoiding quantum woo

Just as we have to be aware that quantum theory is all about models, not 'reality', we also have to be wary of the siren call of an opposing effect. Influenced in part by the vagaries of postmodernism, there has been a tendency to extend quantum observations to the 'macro' world: to think that the Uncertainty Principle, for instance, means that 'everything is uncertain', and that the mysterious nature of quantum theory means that anything that sounds mysterious, and that has a few quantum terms thrown in, has as good a chance of being true as anything else. But that totally misunderstands the situation.

Yes, we are only dealing with models. Yes, it's true that quantum physics does not describe nature, and is only our best way of predicting what will be observed given the current data. But those who play free and loose with the terminology forget that quantum physics is very good at that role. It makes predictions that match reality so accurately that it is the equivalent of guessing the distance from London to New York to the width of a hair. Even when quantum theory says that we can only make predictions based on probabilities because there is no underlying reality, those probabilities are crisp, predictive mathematical values, not vague hand-waving use of terminology.

It isn't the case that all theories are equal in their ability and worth. Nor is it the case that by throwing in words like 'quantum' it's possible to turn fictional woo into anything that even comes close to modelling reality. It doesn't take a lot of searching online before you find 'quantum' devices to magically transform water to 'restore a special balance needed to hydrate' or that bring in special 'energy fields' or other legitimate physical terms and sprinkle them through advertising to give a product a sense of being scientifically based.

The new reality

Simply using the terminology of science, and specifically quantum physics, does not make something valid or useful. But in a way it's not surprising that it is quantum terminology that is most used in this fashion – for the simple reason that quantum physics is so important to our everyday lives. Just as some societies have in the past attempted to copy the outward appearance of technological societies in so-called cargo cults, building mock versions of the real things, so this misapplication of quantum terminology is what Richard Feynman called cargo cult science. It is rubbish in itself – but it emphasises the significance quantum physics has come to have in all our lives.

Welcome to the Quantum Age.

Index